65 Houseplants
From
Seeds, Pits & Kernels

65 Houseplants From Seeds, Pits & Kernels

By Ralf Efraimsson

Published by
Woodbridge Press Publishing Company
Santa Barbara, California 93111

Published and distributed by

Woodbridge Press Publishing Company
Post Office Box 6189
Santa Barbara, California 93111

*Published by arrangement with ICA Publishing House Västerås, Sweden.
Original title: Sätt en kärna!*

Original Copyright © 1976 by ICA-förlaget AB, Västerås

English translation by SCITRAN, Santa Barbara, California 93108

Additional editing by Gloria Leitner

Horticultural consultation on English translation by Margaret Tipton Wheatly
Drawings by Anna-Greta Paulsen
Cover photograph by Studio Svante Sjostedt AB

Library of Congress Cataloging in Publication Data

Efraimsson, Ralf.
    65 houseplants from seeds, pits & kernels.

    Translation of Sätt en kärna!
    Bibliography: p.
    Includes index.
    1. House plants. 2. Plant propagation. 3. Seeds.

I. Title.
SB419.E3513   1977   635.9'65   77-7173
ISBN 0-912800-40-2

Published simultaneously in the United States and Canada

Printed in the United States of America

Contents

Part I — The Basics

Part II — From Seed to Houseplant

Plants From Ordinary Seeds, Pits, and Kernels

We're all used to buying an orange at the store, eating it, and discarding the pits without a second thought. But did you realize that inside one of those lifeless-looking pits is a potential plant?

You can take that orange seed, plant it, and grow a miniature orange tree inside your house—even if you don't live where it's hot all year round. It not only makes a beautiful houseplant, but may actually bear fruit indoors.

You can use the seeds from many fruits to grow tropical and subtropical plants indoors. They may not all produce fruit, but they will develop lovely foliage and often flowers. Picture a miniature avocado, lemon, or papaya tree growing in your own window!

Or think about growing a fig, pomegranate, or kumquat tree, a coffee bean or pineapple plant.

How about pits from the more common fruits we eat—apples, cherries, plums, or peaches? They, too, can be planted and grown as houseplants.

You can also plant certain kinds of nuts, like walnuts, almonds, pecans, peanuts—even coconuts.

Some vegetables do quite well as houseplants, for example corn, peppers, and tomatoes. You can even try growing grains such as rice and barley.

This is a guide to growing houseplants from seeds and pits. Of course, not all pits and seeds can just be popped in a pot with soil. Some are not viable; others require special care and patience. But your energy and time will bring forth rewards manyfold as you see the beginnings of an olive tree or a grape vine or a lime tree adorning your windowsill.

With luck, trees that are natural to your area may eventually be transplanted outdoors. Perhaps the piece of fruit that you buy at the grocery store for a few cents will someday stand as a fragrant apricot or pear tree in your yard.

In these pages you will find specific instructions to follow to produce many different types of houseplants. There is information on how to obtain the seeds, how to know whether they can be planted, what conditions they need to sprout, and how to care for the plants as they grow and mature.

The author also provides a brief history of where the plant originated and how it spread to various parts of the world, how the fruit or nut acquired its name, and where it is grown today. There are illustrated descriptions of what each plant's foliage, flowers, and fruit look like. This will give you a clear picture of what you can look forward to from that inanimate pit you were about to throw away.

Some of the plants discussed here can be grown by children; they are easy to plant and quick to sprout. Growing a carrot top or a bean or an onion can be an exciting first "gardening" project for a youngster. Your children will discover that the food they eat did not magically appear in plastic bags at the grocery store, but came from a living plant.

The author begins the book with a discussion of the

basics—what is a seed or pit, what are the parts of the growing plant, and what types of lighting, soil, watering, and fertilizing conditions will favor a plant's growth. There is also advice on how to deal with pests and plant diseases. This will provide you with a solid factual framework before your green thumb digs in.

Hopefully, this book will also encourage you to try some of the more exotic and delicious tropical fruits with which you may not be familiar—such as the mango, guava, or cherimoya. They will produce some exotic and very pretty houseplants.

In an era of over-sophistication, of being accustomed to buying houseplants from a florist or nursery—already grown and ready to go—it is easy to miss the joy of watching life emerge from its tiny origins. We forget that the dormant seed in the fruit we eat contains the power of life, of bringing forth the slender shoot that will culminate in a fully-leaved, flowering, and lusciously fruiting plant.

To help you rediscover that miracle, to enable you to create a beautiful and unusual variety of houseplants, is the aim of this book.

Gloria Leitner

PART I
THE BASICS

Introduction

According to old Scandinavian lore, it was thought that the whole world lay in a great ash tree, Yggdrasil; and that the first people, Ask and Embla, were created from two twigs, into which the god Odin blew life. To the first people of the North, vegetation meant a great deal. The ash was long regarded as a holy tree, and still is planted on farms in Sweden today, as a protective windbreak.

In many old-country villages, an alderman was elected by the inhabitants, and it was his job to decide on which day the sowing of the valuable seeds would occur. He did this by weighing the seed in his hands and letting it run through his fingers and smelling it to see if it was ready for the earth, and also by touching and smelling the earth to see if it was ready for the seed. This was a very important duty that required much experience. The seeds' growth meant life and bread for the people, while a crop failure meant starvation and death.

Nowadays one year's crop failure can hardly have such devastating effects in the Western world, but we are becoming more and more conscious of the necessity of maintaining the balance of nature and of the fundamental importance of all life on earth. We are becoming aware that the plant kingdom must reproduce and continue to live on. Of the energy Westerners require in order to live, one-fourth comes directly from various kinds of seeds; and for the rest of the world, this figure is as high as three-fourths. Corn, wheat, and rice are the three plants that supply us with most of our seeds or grain.

Many researchers believe that man's first permanent settlements were in places where bananas grew. Instead of always having to search for more fruit, it was soon learned that the shoots of the banana plant would keep bearing. If a person cut down the shoot that had just been harvested, he could hasten the development of a new shoot. Maybe these researchers are correct in believing that tropical fruits have been cultivated for a very long time. Paintings over 3,500 years old in ancient Egyptian tombs show us that even then extensive use was made of fruit. These paintings depict grape growing and tending of the vines, as well as the crushing of grapes for the preparation of wine. In these paintings one can even see people picking figs and pomegranates.

Tropical fruit used to be something that was only heard about in the most northern climates, and rarely seen. It was not so long ago, for example, that oranges were eaten only at Christmas time. Today modern transportation—airplanes and ships specially equipped for fruit transport—has totally changed that, and fruits and vegetables from all over the world are available everywhere practically the whole year.

This availability of many food varieties, plus an increased interest in everything that grows, has resulted in many people attempting to plant the seeds and pits from fruits to see if they can make their own little orange, kiwi, or mango tree grow at home in the window or on a porch or balcony. Even beans, dates, nuts, corn, and peas are planted in more or less successful attempts at cultivation.

The requirements and the conditions of growth for these many kinds of seeds and pits vary. This book is an aid for those who enjoy experimenting with the cultivation of everyday seeds, pits, and kernels. It can hopefully assist in your growing a collection of attractive houseplants which, after a while, will give you a chance to harvest the fruit of your labor.

It may of interest to some readers to know that many of the plants from common fruit seeds can be trained as highly ornamental bonsai or dwarf trees.

What Is a Plant and How Does It Grow?

A plant consists of roots, stems, leaves, and often flowers and fruit. All of these are made of cells that form tissue and organs. Each cell consists of a nucleus, cytoplasm (the surrounding cell contents) , and a cell wall. Cells also contain substances with important functions—such as chlorophyll, which gives color to the green parts of the plant. With the help of chlorophyll and the sun's energy, the plant can transform water and carbon dioxide into simple sugars that can later be used as an energy source and to produce other material for its growth. At the same time, oxygen is given off. This process is called photosynthesis, and has a fundamental significance for all life on earth.

The roots give the plant a foothold in the soil and take up water and nutrients from the earth. The roots also produce certain acidic particles that help to break down the minerals in the earth, which the plant can later use for its own nourishment. Nutrients must always be dissolved in water to be utilized by the plant. Because roots need oxygen in order to function, air must be able to penetrate into the soil. Roots can also store nutrients for a while, but constant overfeeding of a potted plant will bring about damage to the roots and leaves.

The stem's function is to transport water and nutrients from the roots up to the leaves and to carry nutrients down to the roots. In some plants, the stem must also store water and nutrients. In most cases the stem supports the plant, but there are many climbing plants with weak stems that reach for the light by climbing or clinging to other plants or objects.

There are many methods of climbing. Many plants have special climbing-hairs; others have tendrils that wrap themselves around some support or have barbs that attach themselves to an object. There are also plants that grow many small roots on the stem, which then anchor themselves to the soil bed.

Leaves capture the valuable energy from the sun and thus are often especially rich in chlorophyll. Sometimes one can see potted plants in the window turn their leaves towards the light. Many plants and trees have developed ingenious systems of getting their leaves into the light as much as possible. A birch can have a total leaf surface of 400 square meters (4,304 square feet) ! Each leaf is equipped with many "pores," or stomata, that can open and close. With the help of these "pores," that plant can control evaporation and can also absorb carbon dioxide from the air and emit oxygen. The manufacture of simple sugars occurs in the leaf. Plants that come from areas with a rich and regular water supply often have large leaves for the necessary evaporation of water. Plants that come from places with an irregular or sparse water supply often have small leaves or sometimes no leaves at all, like cacti. In cacti, photosynthesis occurs only in the stem and growth is slow.

The flower's function is to attract insects to it by way of scent, color, or form, so that reproduction can occur (although in many plants reproduction occurs in other ways) . For example, the orchid family has flowers that resemble a female insect, and a male insect flies from flower to flower thinking that they are actually female insects, thus transporting the pollen which enables reproduction to occur.

For some plants there have been devastating consequences as the pesticides of modern agriculture have begun to be used. If the pollinating species of insect dies out as a result of pesticides, the species of plant will also die out, since another insect would hardly be interested in or be able to pollinate the flowers. However, the pollination of most plants works quite effectively, resulting in fruit and seeds. In addition, of course, pollination may also be accomplished by hand.

Fruits and seeds provide for the plant's survival, and the dissemination of seeds is achieved in many different ways. One has to marvel at nature's abundant ingenuity in providing a seed with ability to compete for space, so it can develop into a new plant. Many seeds can be spread by streams. Others are equipped with "wings," or with some mechanism that enables them to travel far through the air. Some seeds have barbs that allow them to cling to the coats of animals or clothing of people, and thus be scattered far from the mother plant.

The fruit that grows around a seed is often itself instrumental in the distribution of the seed. When animals eat fruit, the seeds are also consumed, and they end up later in the animals' droppings far away from the place where the plant grew.

Plants can be propagated in other ways too, such as by means of underground stems, bulbs, or tubers.

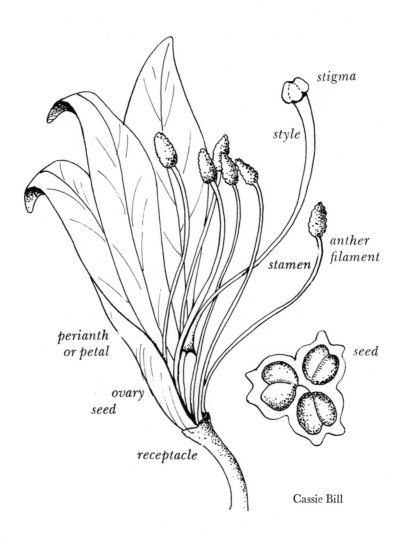

stigma

style

anther
filament

stamen

perianth
or petal

ovary
seed

seed

receptacle

Cassie Bill

What Is a Seed or Pit?

Every flower has, among other things, one or more pistils and several stamen. The underside of the pistil is swollen and is called the ovary. In the ovary there are one or more seeds. Fertilization or pollination can be described in the following manner. The pollen grain that contains the male cells is carried from the stamen by an insect or by some other means from one flower to another. The pollen fastens itself to the pistil's upper portion, which is called the stigma. The pollen grows there and a pollen tube pushes through the stigma and continues growing down through the pistil's style. Fertilization occurs when the male pollen cells reach the female cells in the ovules. Growth of the ovules and seeds begins at that time.

Lingonberries, blueberries, currants, tomatoes, and gooseberries are called berries and contain many seeds. Cherries, apricots, and peaches are called stone fruits since each seed lies protected by a hard "stone." Raspberries and blackberries are examples of multiple-seeded fruits. Pineapple is an example of a multiple-skin fruit. Many fruits have dry, hard walls and if these open and release the seeds when they have matured, one calls the fruit a pericarp. If the fruit does not open and release the seed, the fruit is a nut. A seed can range in size from a coconut to a small begonia seed, of which there are many thousand in merely a gram.

The seed itself consists of a shell and a nucleus. In the nucleus is the germ with one or two rudimentary cotyledon (embryonic leaves) and roots. The nucleus contains enough nourishment for the embryo until the first real leaves have grown, which then enable the plant to produce its own energy The viability of many seeds disappears after several days but there are instances of seeds that have existed for a thousand years and are still viable.

The seed needs three things in order to grow: water, oxygen, and a sufficiently high temperature. When the seedling sprouts up a bit, it also needs light. Water is needed to soften the shell so that it will be easier for the sprout to break through; it is also needed so that the nucleus can swell and break the shell. Oxygen is required in order for the life processes to begin in the sprout. Therefore, air must be available to the seed.

For many seeds, a temperature of at least 10–15°C. (50–59°F.) is necessary before anything can happen. For seeds from tropical plants, the temperature must be 18–27°C. (64–81°F.) before they will grow. When the sprout with the cotyledon comes forth and later the first real leaves have grown, the plant must also have light to be able to produce its own energy.

Many seeds have an internal biological clock which dictates that they cannot grow before they have been exposed to a low temperate parts of the world, where the seeds must wait until spring comes before they grow. This process is called wintering, and it can occur by artificial means, which we will talk about later. Seeds can also be equipped with a practically invulnerable shell to prevent water from seeping in, and these seeds often come from areas with considerable rainfall. In this case one can nick the shell with a file or thin it with sandpaper, and thus help to hasten the sprouting process.

Besides those conditions which have just been described, there are various other environmental factors that must be considered if the seed is to sprout and grow, and a discussion of these will be found later on in the book.

It is perhaps needless to say it, but you should, of course, select seed from fruits or vegetables that are uncooked and not treated in other ways that would destroy the seeds' viability.

Pots and Cultivation Boxes

To begin with, you must have something to set the seeds or pits into, and there are a number of cultivating containers available. Old-fashioned clay pots undeniably have many advantages, perhaps most importantly their ability to "breathe" and let air and water through to the roots. It is vital that oxygen be able to get down into the earth, and therefore the soil in your pots should never get too wet. Plastic pots, cut-off cartons with holes in the bottom, and other containers can be used, but since they do not "breathe" one always runs the risk that the earth will become too wet. On the other hand, these types of pots have the advantage of keeping the soil slightly warmer—but if you place a clay pot in a warm place, the temperature of the soil should be adequately high.

Lately a number of special containers and mini-hothouses have appeared on the market. These have many advantages when you cultivate seeds or pits. They have, as a rule, a transparent lid that lets in light and allows the maintenance of a high level of humidity, which is a necessary condition for success with most of the plants described in this book. Furthermore, there is room for many seeds in the same container, and the container can be placed where it gets heat from below, which increases the soil's temperature. Since the humidity of the soil is kept higher in these containers, you need not water more than every 10–14 days if you spray the surface of the soil every day. Watering only at long intervals is very good for the oxygen content of the soil.

When planting in a regular clay pot, invert a drinking glass over the soil to increase the humidity; or set the pot in a plastic bag cutting out several air holes. If it becomes too humid in the plastic bag or in the container, air it out. When planting in this type of pot, place sand or coarse gravel in the bottom of the pot to make it easier for water to drain off and for oxygen to penetrate into the soil.

Another good device you can use is a pressed peat pot (Jiffy pot) which swells up into a small net-like pot when watered. Peat pots can be placed in a container, and each pot planted with a seed

Jiffy-pots expand in water and make good starting containers. Can be planted directly into the ground or a larger container.

or pit. Using this method, you can avoid transplanting when the plant has grown larger. Simply transfer the entire peat clump into a large pot with ordinary plant soil. Thus, you will not have to touch the root system unnecessarily, and the plant will not be disturbed.

In the beginning, before the plant has fully grown up, it is best to water from underneath the pot. Only the daily spraying of the soil should be done from the top. There are many types of spray bottles you can use that deliver a fine spray. Make sure that the soil is not too wet, because then the seeds or pits will rot.

Soil and Fertilizer

The soil's most important task is to provide something for the plant's roots to hold onto and to contain the water and nutrients that the plant requires. The soil structure must be porous, so that air can get into the soil for the roots to breathe.

Soil composition can vary quite a bit, from garden soil to pure clay to mulch. Packaged soils also vary greatly in composition, but most are rich in mulch with varying proportions of other soils. In addition, they are enriched with various minerals. These prepared soils are good for growing plants from seeds, pits, and kernels.

If you have access to soil from flower beds, it is probably good to use; though you always run a certain risk of pests that you avoid by using purchased soil—which is sterilized with steam. The soil in containers or peat pots is also good for seed cultivation.

Of the sixteen basic substances that a plant requires, thirteen are found in the soil. The most important of these are nitrogen, phosphorus, and potassium; but lime is also one of the required minerals. In addition, the plant needs a number of trace minerals, such as iron, copper, boron, etc.

As a rule, the soil contains adequate amounts of these basic nutrients for the plant's needs during the first months after sowing the seed; but after that fertilizer is necessary. When you plant seeds or pits, you should be very conservative with fertilizers, but you should provide a small amount as soon as the first real leaves are developed on the seedling. Later, as the plant grows larger, increase the amount of fertilizer, never giving more than what is recommended on the package of the preparation used. In the winter, most plants need to be fertilized only once. For most of the plants described in this book, use the same vitamin and fertilizer preparations used for other houseplants.

Water and Air

Water and air have a great influence on the seedling's well-being after it has come up and developed its first real leaves.

Fresh air is necessary for the seedling's development and health. Many seedlings are very sensitive to pollutants such as exhausts, smoke, and the influence of ethylene gas from ripening fruit, as well as gases from the building material of new houses. It is good to air out a room where there are plants, but remember that for tropical plants a temperature of even 4°C. (39°F.) will result in permanent damage. These plants ought to be moved before you ventilate a room in cold weather.

Besides the cold, these plants do not tolerate drafts. A little banana plant, for example, will wilt very quickly if it stands in a draft. The air should be rather humid. In newer houses the air is usually dry, but this can be helped by daily spraying the leaves with lukewarm water. For small plants in mini-hothouses or those with a plastic bag around the pot, humidity is no problem.

Plants are dependent upon fresh water, especially tropical and subtropical plants. Some tap water cannot be used unless it is boiled first and then allowed to cool. This procedure should be followed if there is a high chlorine content or too much lime in the water, which are poisonous to a number of plants. If you have access to rainwater, this is perfect. Always use water at room temperature of lukewarm water.

Planting and Placement

On the seed packages available in the stores there are often detailed instructions for planting. With seeds or pits from fruits it is not quite as easy to know how to proceed since the various seeds should be planted in very different ways, so the procedures will be described carefully for each type of plant later in the book.

The first factor to consider in deciding where to put the pot or container is the availability of light. When the plant has grown its first leaves it should be placed, generally, somewhere with ample light but no direct sun. The temperature of the soil should average between 18–27°C. (64–81°F.) for the seeds to grow, so a source of warmth from beneath the pot may be necessary. You can do this by simply putting the pot under a window with a heater beneath it; but if the windowsill is marble this will chill the soil, so you will have to provide warmth in another way. One method of doing this is to place the plant on a table in the vicinity of the window and the heater.

The temperature of the soil can most easily be checked by using a regular outdoor thermometer.

How Long Will It Take To Grow?

How long before the seed or pit will grow is a question of great interest to the person who has planted and is eagerly waiting for something to happen. It is impossible to generalize since so many factors play a part, and also seeds from the same sort of plant can vary considerably in the length of time they take to grow.

An avocado pit, for example, can require between 3–15 weeks before a shoot comes up. With other pits, "wintering" must occur before anything happens. How this wintering takes place is described on page 44. Dates can take quite a long time before they are ready to grow, and only then should they be planted. If you are going to cultivate a date palm from a pit, you must be equipped with great patience. Other seeds and pits can grow very quickly, and for those who are not patient many of these plants are described in this book.

Even if you follow all of the instructions outlined in these pages, some seeds and pits will not sprout. Modern methods of agriculture throughout the world aim, naturally, to steadily increase the yields of plants. This has resulted in some varieties of trees with very good, but unfortunately sterile, fruit—the seeds cannot produce a new plant. Such fruits are produced commercially by grafting or budding.

With the stone fruits, the nucleus may be missing from the pit—which you cannot detect by looking at it but will certainly discover after a while when nothing happens after it is planted. There is not much you can do about this, except to summon up fresh courage and plant again. Most seeds and pits will, after a while, result in a seedling and you need not be discouraged, even if it takes a long time for something to happen.

Transplanting

When the seedling has grown and needs more room for further development, the first transplanting is necessary. This should occur rather soon if you have sown in a flat—a shallow growing tray. For the first transplanting you may put several seedlings in a larger pot, or plant one seedling in a small pressed peat pot. The plant's roots reach through the peat pot after a while. Then you can plant the entire peat pot into a larger pot with regular potting soil.

The soil in the new pot should be well moistened when transplanting. Make a hole in the soil with a pencil where the seedling is to be placed. When picking the plant up out of the flat, be sure not to damage the roots. Try to lift the plant carefully with your fingers. A small tool (even a pot label) can be used to bring the soil along with the roots.

We mentioned earlier that seedlings planted in peat pots do not need to be transplanted but can continue to grow in those pots for some time. However, when the plant is between 10–15 cm (4–6 in.) high, it should be planted in a larger pot with regular potting soil. Simply lift the entire clump of peat into the larger pot. The advantage of this is that you can avoid injuring the root system.

After a while when the plants have grown, transplant again. Tropical plants require relatively large pots to do well (one exception to this is the pineapple).

Larger plants are transplanted at several-year intervals, but it is best to refill the pot with new soil every year. For older plants, cut away some of the old roots and pick away some of the old soil. In this manner, the same pot can be used with new soil.

Topping and Pruning

For many of the small plants it is important that early and repeated topping be done. Otherwise, there will be only one long and lanky shoot, instead of a bushy and richly branched plant. The first trimming ought to occur when the little shoot is about 10 cm (4 in.) high. If you are lucky the plant will branch, and when these branches are about 10–15 cm (4–6 in.) it is time for another trim, etc. The time for trimming will be discussed for each plant later in the book.

After the plants are several years old, they should be pruned. Usually this is done in the early spring. Since the growth process for the majority of these plants is quite slow, prune with care! Fruit-bearing plants can be pruned after the harvest of the fruit.

Orange trees, for example, can be pruned branch after branch as the fruit ripens and is picked.

Many tropical plants need a great deal of light, and it can be difficult for them to survive during the darker days of winter in northern regions. By putting the plants in a very light place near a southern-exposure window, they can obtain an acceptable amount of light. If the remaining requirements—such as warmth, humidity, and watering—are met, the winter will not be any problem for these plants. However, if it is not possible to place the plants in a spot with enough light, it may be necessary to arrange some kind of artificial lighting so that they will have a chance to survive.

Artificial Lighting

Artificial lighting can be set up in several ways. The best alternative is to use the so-called mercury or mixed-light lamps. However, these lamps give a rather strong and blinding light that is not especially comfortable in our homes. Furthermore, the installation of special electrical fittings is required, which is both costly and troublesome. In the home environment, a combination of fluorescent tubes that are designed for plant lighting along with regular electric bulbs is the best alternative. Fluorescent lights are easily softened by simultaneously using electric bulbs, preferably located in several spots. This will give the plant effective lighting. In order for the light to give the plants energy, the fluorescent lamps should be rather close to the plants, 8–10 cm (3–4 in.) away.

To provide seeds and plants with extra lighting, use regular

75-watt electric bulbs that are placed about 20–25 cm (8–10 in.) above the pots or flats. For the full effect to continue throughout the winter, the lights should be on as many as 14–16 hours per day if necessary. In the spring, as the natural light increases, successively diminish the hours you use the artificial lighting.

As the plants grow, gradually raise the light source so that it is always well above the plants since it is easy to get burn damages from the heat of the light. Also, remember that small plants getting artificial lighting need more water than usual.

Will There Eventually Be Fruit?

The main reason for cultivating tropical fruit plants is not so that you can eventually grow a large supply of fruit—even though that would be desirable—but because you can produce a collection of beautiful plants to put in your windows. Green plants have become a very popular addition to the home environment, but some are quite expensive. The hobby of cultivating seeds and pits can provide a less expensive alternative.

However, you can still dream of being able to pick some oranges from a tree you have cultivated yourself! Many plants can indeed produce fruit if you help them along by taking certain steps such as trimming and pruning the plants, as well as pollinating them when the flowers have blossomed.

For each fruit-bearing plant described later in the book there are instructions telling you how you can help them to bear fruit.

'Wintering' Seeds

As mentioned earlier, some seeds will not grow unless they are first exposed for a time to a temperature of 1–6° C. (34–43° F.). These seeds have their natural growing seasons in their native habitats, and must therefore wait until the spring before they grow.

At home you can arrange this wintering process by setting seeds or pits in sand or peat moss, which will maintain humidity. You can keep the seeds either outside or, if you prefer, in the refrigerator. If you put them outside, they can be exposed to freezing temperatures a bit; but if it gets too cold, insulate them. In general, the seeds should stay cold for about three months.

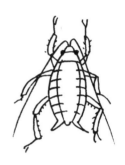

Pests and Plant Diseases

Unfortunately, all plants are occasionally attacked by pests and other plant diseases. As a rule, it is hard to discover how the plant has become infected. Plant diseases can be spread by the wind and thus can even be blown in through an open window. A bouquet of wild flowers is often infested with pests, easily spread to houseplants. If you discover the problem early, it is usually easy to solve it. However, plant diseases such as those caused by viruses and bacteria, which are spread by pests, are difficult or even impossible to cure.

When you suspect that a plant is infected, you should isolate it immediately so that the pest does not spread. Then find out which type of pest it is, since the methods of fighting each type can be quite different and not all of them can be eradicated by aerosol sprays.

Aphids are the most common pest and multiply very quickly. Aphids are usually green and sit on the tops of new shoots and on the flower buds—which will wither rapidly. They can be fought outdoors with malathion; or, indoors, with pyrethrum or rotenone preparations. Aphids often spread viruses and bacteria as they attach themselves to the leaves, and this becomes evident in the form of spots that slowly spread and cause the entire leaf to wither. Infected plants must be thrown away.

Red spider mites appear on the leaves' undersides, where they spin a mesh of small threads barely visible to the naked eye. Mites travel fast in sunny and warm weather, and quickly spread from plant to plant. They are very small and difficult to detect, but if they are discovered early they can be eradicated by removing their webs from the leaves with cool tap water (put the plant under the running tap). If the mites have spread, dry off and spray the leaves with an oil emulsifier. Repeat the treatment according to the directions on the preparation.

Thrips are insects somewhat larger than 1 mm, with long fringed wings. Infected leaves acquire a dull silver color and new leaves are dwarfed. Eradicate as described above for aphids.

White flies and many other small insects that live in the soil can be fought by immersing the pot in cold water. When the animals and eggs float up to the water's surface, strain them off. After such a treatment, it is important to give the soil a chance to dry thoroughly before you water again. The fact that these pests do well and develop can be a sign that the soil is always too wet.

Mealybug is a common pest of many of our houseplants. Oranges are especially prone to infection. It is easy to discover mealybugs, with their characteristic brown shields under which the females lay many eggs. You can combat them by first scraping away the shields and then brushing the leaves with alcohol, using a small cotton swab.

Cottony cushion scale is a close relative of the mealybug, but it secretes many wool-like threads that become twisted in the leaf joints and other nooks of plants. It is rather difficult to fight this pest, especially if it has been developing for a while. Try to scrape away their "wool" and then brush the plant with alcohol, as described above for mealybugs.

To fight all pests, carefully follow the directions on the package of the respective preparations. The preparations may be applied with a plant mister. If you use aerosols be sure the preparation is formulated for plants and also be sure to hold the can a distance from the leaves when spraying. Otherwise, the leaves can be damaged by the cold gas emitted from the aerosol can.

For lighter cases of red spider mite, aphids, and thrips, mix a solution of 100 parts water to 2 parts dishwashing detergent, and spray it on the plants.

PART II
FROM SEED TO
HOUSEPLANT

Almond

Prunus amygdalus

The almond tree belongs to the *Rosaceae* family. It is a close relative of the cherry, peach, and plum. The almond came originally from Persia and Afghanistan, but it spread early both west and east and quickly became quite common in countries around the Mediterranean. On the island of Mallorca many trees were planted in the middle of the eighteenth century, and today there are over six million almond trees there. They are also abundant in some other parts of the world.

The blossoms of the almond tree, which emerge in the early spring, have been the subject of poets' songs throughout history. Almond blossoms adorning the trees at the height of the season are an unforgettable sight. The blossoms have practically no stems and sit directly on the branches. The flowers are white, while the buds are pink. The leaves remind one of the peach. The fruit is downy and greyish-green and is considered a stone fruit. Inside the fruit is the seed, surrounded by a very hard shell in the sweet and bitter almond varieties, while the type called the dessert or soft-shelled almond has a soft shell. Dwarf almond trees, *P. tenella*, are grown only as decorative garden plants.

For the planting of almonds, see Apricot, page 55.

Apple

Malus domestica

The apple is a member of the *Rosaceae* family, and the wild apple is found in many parts of Europe as far north as central Sweden. Crosses between different wild types are thought to have resulted in the cultivated trees we have today. The apple is our most important fruit tree and its fruit has been a favorite for a very long time. Studies have shown that apples were eaten in various areas even during the Stone Age. In southern Europe, apples have been eaten and cultivated for thousands of years, and the apple has been used as a symbol of all fruit. It is commonly alleged that an apple tree was the tree of knowledge from which Adam and Eve ate, and the apple has been used as a symbolic design in the orb of the British royalty's regalia.

Today there are more than a thousand cultivated types of apples. A great deal of research is continually going on around the world to develop new varieties of apples with improved characteristics.

The apple is not produced commercially from seed, but by grafting. The chance of getting the same type of fruit from a seed-grown plant is one in a million. The blossoms of some varieties of apple trees—like those of many other fruit trees—are self-sterile. This means that only pollen from another tree can fertilize them. There must be at least two trees near each other for one to bear fruit. The pollination of fruit trees is accomplished mainly by bees. There are some newer kinds of apples that can produce without being pollinated. However, the seeds of these varieties are sterile and cannot reproduce a new plant.

The apple is like the pear, in that both fruits are created from an enlargement partly of the pistil and partly of the surrounding flower shaft. The core of the apple has five cavities, with two seeds in each cavity.

Apple seeds should be wintered before sowing, and this can be done in regular potting soil that has been well-moistened (see page 31) Then plant the seeds in clay pots or flats, or in pressed peat pots (Jiffy pots) that swell up when watered. Bury the seeds one cm (less than half an inch) . A plastic bag placed over the pot, with air holes in it, or a lid over the flat will increase the humidity. Put the pot in a light spot. with average room temperature, and spray the surface soil daily. Thus, you will need to water only at long intervals, and this will allow time for air to penetrate down into the soil.

When the plant has come up and grown a bit, it should be transplanted into a larger pot. Spray the leaves every morning, using a sprayer that finely distributes the water. Water periodically and fertilize regularly.

Early in the summer, plant the tree in the garden or in nature where it can acclimatize itself. One cannot grow an apple tree indoors for any longer period of time.

Apricot

Prunus armeniaca

The apricot belongs to the *Rosaceae* family and is a close relative of the plum and cherry. The plant came originally from Turkestan. It had been grown in China long before it was first brought to Italy from Armenia around the time of Christ's birth. (Armenia lies at the border between present-day Turkey and Russia.) The fruit derives its species name (*armeniaca*) from Armenia, and the name apricot comes from the Latin *praecox,* which means early ripening.

The apricot is a stone fruit, which means that the outer part of the fruit is meaty—the part we eat—and inside is the "stone," or pit. In each pit or stone there is usually only one seed.

Apricots are often sold dried, without pits, but they are also available fresh during the summer in most areas, thanks to modern transportation. Their taste is reminiscent of the peach, a close relative. The skin is reddish gold and slightly downy. The pulp of the fruit is orange and the pit is brown and flat, with sharp edges on one side.

Apricot trees can grow as high as 3–8 m (10–27 ft.) . The leaves are somewhat heart-shaped and the blossoms are white with a red underside.

The apricot pit has a very hard shell to protect the seed from water. Before planting it, file or sandpaper the pit so that the shell is noticeably thinner. Then plant the pit in a small pot with regular potting soil or pure mulch that has been moistened. The pit's point should be covered with soil.

The soil should be kept consistently humid—placing a glass cover or plastic bag, with air holes, over the pot will provide high humidity. This will make possible long intervals between watering. Spray the soil surface daily so that the seed will grow quickly.

When the plant has grown about 10 cm (4 in.) , plant it in a larger pot with regular soil. Put the pot in a very well-lit place with protection against the strong sun. Water at intervals, so that in between the waterings air can reach down into the soil. Fertilize the plant during the warmer seasons. Spray the plant daily with water.

It is doubtful that flowers and fruit will develop indoors. In some areas, when the tree has grown larger you may plant it outside in a protected spot.

Avocado

Persea americana

The avocado belongs to the *Lauraceae* family, which also
includes the laurel tree. The name avocado comes from the South
American Indian name for the fruit. One species of avocado
comes from Guatemala and the West Indies, while the thin-
shelled type, *P. drymifolia,* comes from Mexico. Modern varieties
are crosses between the original types.

The avocado has become more common in recent years. The
taste is best when the fruit is quite ripe, and this is also the best
time to plant the pit. Varieties with rough skins have given the
avocado the name "alligator pear," and these types have pits that
are easily grown.

Planting avocado pits can be done in several different ways,
but the best method is the following. Use a small pot with regular
potting soil or pure peat. After the pit is taken out of the fruit, it
should be thoroughly washed in lukewarm water and dried. Then
it should lie in a warm, dry place for a day, so that the brown skin

surrounding the pit dries and cracks. Plant the pit so that two-thirds of it lies under the soil and one-third is above the surface. The pit will sprout quickly if you cut several deep incisions in the top and bottom of the pit with a sharp knife (the larger end is the top).

Always moisten the soil before planting; and after the pit is planted, invert a glass over the pot or tie a plastic bag, with air holes, around it. This provides a high level of humidity and allows for long intervals between watering, so that air has a chance to penetrate down into the soil. Spray the soil surface with warm water daily. It can take 3–15 weeks before the seed sprouts, but then the plant will grow rather quickly. When this occurs, fill the pot with more soil so that the pit is covered.

The pot should sit in a light and sunny place, but the root system is prone to getting dried out, so it is good to have an outer pot as protection from the sun. Sand or coarse gravel in the bottom of the pot will keep the soil moist. This provides a certain amount of humidity around the plant, but it is still advisable to spray the leaves daily with water that is at room temperature.

It is important that the avocado plant be topped early so that it will spread out and become a beautiful home ornament. When it is about 25 cm (10 in.) high, cut off 7 cm (3 in.) from the top. When the new shoot has grown a little, trim this also, and so on. The soil should be kept humid the entire year, and vitamins or fertilizers should be applied during the warmer months. With luck, the plant will blossom in several years, but you may not produce fruit indoors even by inducing pollination by transferring the pollen from flower to flower with a brush.

Another way to plant an avocado pit is to make several holes on opposite sides of the pit. Put toothpicks in these holes and then hang the pit over a glass. Put water in the glass so that it covers at least half of the pit. In this way, you can observe the development of the roots. When the plant's shoot first appears, plant the pit in a pot with regular soil.

Banana

Musa paradisiaca

Banana plants comprise a single family, *Musaceae* in Latin. There are about ten wild types. The name derives from Kaiser Augustus' doctor, whose name was Musa. The banana is a many-yeared giant plant, which can reach up to 10 m (33 ft.) high. It is possibly the oldest cultivated plant. Some researchers believe that man's first permanent communities existed in areas where wild bananas grew. These people harvested the bananas, cut the plant down, and saw that a side shoot would produce a new harvest.

Europeans first came into contact with bananas because of Alexander the Great's campaign to India in 327 B.C. The Arabs carried the banana to Africa, and in the Middle Ages the first bananas came to the Canary Islands. From there they were carried by the Spaniards to the Americas, and today about 85 percent of all bananas are exported from Central and South America.

Plantains also belong to the banana family, and they are quite common in Latin America. These green fruits are boiled or roasted before eating. Common bananas are crosses of the various wild types, and new sorts are continually being developed. Most edible types of bananas have lost the ability to produce seeds, and can only be reproduced from side shoots. The small brown seed-like particles in bananas are rudiments of what at one time were seeds.

Some of the less edible bananas still have the ability to develop seeds that can reproduce new plants. Sometimes you can find these sold in the stores; they are the varieties called *Musa coccinea, Musa arnoldiana,* and *Musa ensate.* The banana plant can be a pretty potted plant in your window, but you cannot expect to obtain any edible fruit.

Before the seeds are to be planted, they should be nicked with a sharp knife and then soaked in warm water for about 24 hours. Plant the seeds in peat or regular potting soil, several in each pot of flat. The soil should be well-moistened. A plastic bag, with air holes, or a glass top over the pot will raise the humidity. The soil should be warm, and the pot placed preferably near a heater. Daily spraying of the surface soil means that watering will only be necessary at long intervals.

It can take one week to four months before the seed sprouts. When the seedling comes up, it is placed in a pot near warmth and light (but not direct sunlight) . When the plant outgrows the first pot, it should be transplanted into a larger one.

In summer the banana should have a great deal of light, and it tolerates sun, but remember that the leaves will be easily burned if the plant stands in front of a window with a southern exposure. The soil should never be allowed to dry out, and therefore it is a good idea to have some kind of outer pot. The soil should be kept consistently humid during the summer, and watered less often in the winter. Use a rich nutrient preparation during the warm seasons.

It is important that the banana plant grow in high humidity, so daily spraying of the leaves is necessary. Once in a while, set the plant in the bathtub, and spray it thoroughly with warm water so that dust and other particles that collect on the leaves are removed.

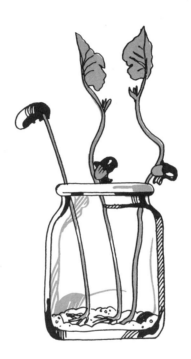

Beans

Phaseolus vulgaris,
Phaseolus coccineus,
Phaseolus lunatus

There are many types of beans, some of which are used in cooking. Beans belong to the *Leguminosae* family.

The garden beans, *Phaseolus vulgaris,* include French string beans, climbing beans, and wax beans. Their entire pod is used in cooking, while brown and white beans, which are also considered garden beans, are used as peas or seeds.

The scarlet runner bean, *Phaseolus coccineus,* is raised mostly as an ornamental plant, but its pods can be used like string beans for cooking, and their beautifully marbled seeds are also edible. Their flowers, white and red or all red, resemble those of the sweet peas. Lima beans, available in the stores, are called *Phaseolus lunatus* and have large flat seeds.

Beans have been raised since the earliest times by the Indians of South America, where the majority of bean varieties originated.

Beans can be planted in many ways. Because they grow quickly, it is exciting to experiment with them. Place some beans in a glass jar with a layer of wet cotton on the bottom, and then put a lid on the jar. After several days, you can see how the root comes forth and tries to nestle into the cotton, and then the cotyledon emerges, and after a while, the first real leaves. If it gets too damp in the jar, take off the lid and air it out, and if it gets too dry, add water. When the plants have grown a bit, plant them in a large pot with regular soil.

Most beans are climbing plants, so arrange some sort of support for them. They provide a green and pretty frame around a window, and flowers and pods can develop. The best results are achieved when planting seeds in the early spring.

You can also cultivate beans in a pot with slightly moist, regular potting soil. Tie a plastic bag, with air holes, around the pot. When the beans have grown up, the soil should be kept moist and fertilized about once a week.

Bean plants should be placed in a sunny location, with protection against the strongest midday sun. Spray the plants with lukewarm water in the morning and evening. Use a spray device that distributes the water finely. Bean plants live one year, and you can save the pericarp (bean pod) for next year's planting.

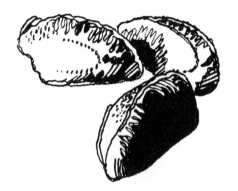

Brazil Nut

Bertholletia excelsa

The Brazil nut belongs to the *Lecythidacea* family and comes from Brazil, where it is a tall jungle tree. Like many other jungle plants, the stem has a sort of buttress that comes out of the juncture between the roots and the trunk. The leaves are smooth, wide oblongs with strong vein marks, and the flowers are creamy white. The fruit is round and large with a stonehard shell, and in this lies the seed or what we call the Brazil nut. The name Para nut, which it is often called, comes from the city Pará (now Belem) in Brazil where much of the exporting of the nuts is done.

The culture of Brazil nuts can be a lengthy process. First, sandpaper or file the shell of the nut so that the sprout will have an easier time breaking through it. Plant the nut in a deep clay pot with light, moistened soil. Bury the nut 2–3 cm (¾–1¼ in.) in the soil. Make sure the bottom of the pot has good drainage for better air circulation in the soil, which will be beneficial. Put the pot in a warm place, preferably over 25°C. (77°F.) . A plastic bag, with air holes, placed over the pot to increase humidity will also encourage the plant to grow. Spray the top soil daily with warm water, and water it thoroughly only at intervals so that air has a chance to penetrate into the soil.

When the plant comes up, put it in a light and sunny place, and water it often, adding ample fertilizer. Daily spraying of the leaves is advised. Transplant the seedling after it has grown somewhat.

Carrots

Daucus carota

Carrots belong to the *Umbelliferae* family and are found wild in Asia, North Africa, and many parts of Europe. Edible carrots are ancient crosses between wild types. Carrots are biennial plants, which gather nourishment during the first year in the edible root.

If you want to experiment and discover the dynamic energy in this root, cut off the upper part of the carrot so that you have a thin slice of about one cm (½ in.) . This should be placed on a plate which is then filled with water reaching about halfway up on the slice. Put the plate in a warm and light spot. After a while, leaves will shoot up, and they will look green and beautiful on your windowsill.

Using this same method, you can grow other root vegetables such as parsnips, beets, and rutabagas.

Cherimoya

Annona cherimola

The cherimoya or custard apple belongs to the *Annonaceae* family and originally came from Ecuador in South America. Now the tree is grown in generally the same places that citrus is grown throughout the world.

This is an uncommon fruit, but it can be found in special stores from midwinter through early spring, at least in the bigger cities. The fruit has a characteristic fish-scale pattern, and it contains many black seeds.

Pick out several seeds and wash them carefully in warm water. Let them dry about 24 hours and then plant them in a pot or flat. It is fine to use regular potting soil. The small pressed peat pots (Jiffy pots) that swell up when watered are also ideal for planting cherimoya seeds. Keep the soil moist, but make sure it is not water-logged. Spray the top soil layer daily with warm water. Put the pot in a warm place.

When the cherimoya comes up, it has to be transplanted into a bigger pot if it was planted originally in a flat or with several seeds to a pot. It should be placed in a spot with a lot of light (but not direct sunlight). Spraying the leaves daily with warm water is essential for the plant's well-being. Water regularly during the summer with plenty of nutrients, and water slightly less often during the winter when the plant is dormant.

You probably will not obtain any fruit from this plant, but it has very decorative leaves and can be a beautiful addition to your collection of houseplants.

Cherry and Cherry Plums

Prunus avium,
Prunus cerasus

Both of these types of cherry belong to the *Rosaceae* family and grow wild in Southwest Asia. Wild cherries that belong to the sweet cherry family grow in some areas of Europe. Sweet cherry trees are quite tall. The most important types are the white-hearted cherries.

Sour cherries are a very ancient cultured plant from the Mediterranean, and they were carried throughout Europe by the monks as they built their monasteries during the spread of Christianity. Sour cherry trees are slightly smaller. In folk medicine, practically all parts of the cherry tree have played an important role—for example, the juice of the bark was used for coughs.

If you are going to plant cherry pits, you must winter them first. Lay them in a pot with lightly moistened sand or moss, and place the pot in a dark location with a temperature of 1–6°C. (34–45°F.) —such as in your refrigerator—for about 2–3 months.

(See page 44.) If the pot is outside, it is alright to let the pits be exposed to frost for several days; but if the frost continues, the pot should be insulated with newspaper.

Before planting in a pot or flat with regular, moistened potting soil, you should file or sandpaper the pit so that the sprout has an easier time breaking through the shell. Bury the pit about one cm (½ in.) under the soil. Place the pot where it is exposed to light and put a plastic bag, with air holes, over the pot or a cover over the flat, so that the humidity is kept high. Spray the surface of the soil daily and water every 10–14 days.

When the plant comes up, spray it daily, and when it needs to have more space, transplant it to a larger pot. Do not keep the cherry tree inside indefinitely; instead, plant it in the garden early in the summer—then it will have a chance to survive. Unfortunately, the cherry will not be a true variety if grown from a cherry pit, and its fruit will not be as meaty.

Citron

Citrus medica

Citron belongs to the *Rutaceae* family and resembles the lemon, except that the fruit is larger. It has a very thick skin and little pulp. Its origin is unclear, but Alexander the Great's soldiers became acquainted with it for the first time in the province of Medica, in what was then Persia, and they called it the "apple from Medica." This was the first citrus fruit to come to Europe.

The fruit's skin is fermented and candied and is then used in cakes and cookies. In many places citron is sold as dried candied peel, but in areas where the tree is grown, you can obtain whole fruits.

For the planting of citron seeds, see Orange, page 109.

Coconut

Cocos nucifera

The coconut palm belongs to the *Palmae* family, and according to some researchers it originally came from Southeast Asia. It is considered a feather palm and can reach 20–25 m (66–83 ft.) tall. The palm especially likes to grow near beaches, and the nuts (which can be carried on the water) are spread over vast areas of the Pacific Ocean.

To succeed in growing a coconut, it is best if you can get a nut that is still in its outer shell. The coconuts that are usually sold in the stores do not have these shells, and they have thus lost their viability. Sometimes, though, you can find them still encased in their outer shells, and these you can plant.

Put the coconut in a large pot or dish filled with sand. Position the nut in the sand so that the wider end is slightly higher than the narrower end. Then all of the juice or milk in the nut will collect at the growing end—the narrower end. The sand should be moist, and every day spray the narrow part of the nut (where the sprout will eventually emerge), as well as the part where the nut was at one time fastened to the tree. In this way, the outer shell will be kept humid, which will hasten growth.

As soon as you discover that the sprout is beginning to emerge, plant the nut in soil. You can use regular potting soil, but mixing in 30 percent sand affords better aeration and drainage. It is important to be sure that the bottom of the pot drains well, because palms are dependent on getting air through the soil. Plant the coconut so that two-thirds is below the soil and one-third is above. Keep the soil moist, but do not let it become too wet.

When the first leaves have come up and divided—so that they actually resemble palm leaves—cover the remainder of the nut with soil. The nut will eventually decay in the pot and turn into soil. During the summer, the palm should stand in a light and sunny place, with protection against the strongest midday sun.

Give it fertilizer during the warmer times of the year. Spray the palm leaves only moderately; but in the winter, due to dryness from heating in the house, it may be necessary to spray them regularly (preferably in the morning).

Coffee Bean

Coffea arabica

Coffee belongs to the *Rubiaceae* family and is native to tropical East Africa, where the plant still grows wild in Ethiopia. The coffee bean comes from a tree that can reach up to 9 m (30 ft.) in height, but it is cultivated as a bush. Cultivation of the coffee plant began in Aragia around Mecca, and during the seventeenth and eighteenth centuries the plant was spread to South America, where the largest crops are now grown.

The coffee bush is relatively easy to cultivate. The leaves are shiny green and come out in pairs. The flowers are small, white, and sweet-smelling. The fruits look like very small cherries and sit in the folds of the leaves. Inside the coffee bean there are usually two seeds surrounded by a thin seed-shell that consists of a parchment-like membrane, and the seeds are situated inside a juicier pulp. The coffee bean is considered a stone fruit.

It can be very difficult to obtain viable coffee beans. Raw coffee beans, which may or may not be viable, may be obtained at some coffee stores. Of course, if you happen to know someone who is already growing a coffee bush and it has fruited, you may use these beans. The viability of the coffee bean is fleeting, so experiment with the freshest seeds you can find.

When the coffee beans are ripe, pick them off the plant and separate them carefully, picking out the seeds. They should be planted in a pot or flat, barely covered with well-moistened soil. The soil should then be kept moist, and daily spraying of the uppermost soil layer permits watering only once in a long while. A plastic bag, with air holes, or a transparent lid over the flat makes the air more humid, which is good for the plant. It is very important that the temperature of the soil be around 24–26°C. (75–79°F.) . The pot should, therefore, be set in a warm place where it is very light but not sunny. Coffee is easily damaged by heat, which affects the leaves.

When the plant is between 10–15 cm (4–6 in.) high, transplant it into a larger pot with regular potting soil. Watering should be regular, but remember that air must be able to reach down into the soil for the roots to breathe. Nourishment should be given during the warmer seasons. Daily spraying of the leaves, preferably in the morning, makes the coffee plant healthy and prevents the leaves from getting brown edges, which otherwise can often occur due to dry air. Some of the leaves may droop or fall but there is not much to be done about this.

After about four years the first blossoms will appear, and these can be artifically pollinated, resulting in fruit formation. It takes about seven months before the fruit is ripe.

76

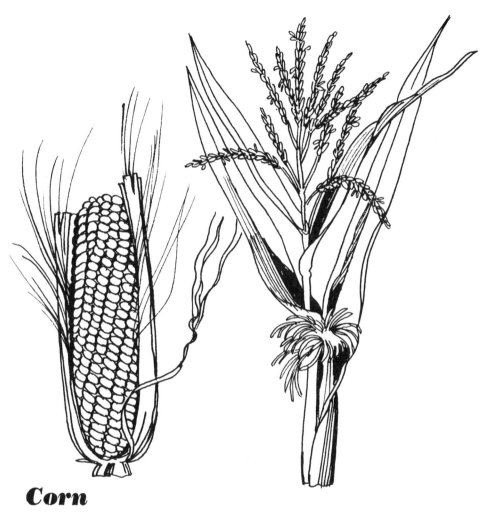

Corn

Zea mays

Corn is actually a grass (the *Gramineae* family). It probably originated in Mexico, where pollen grains have been found from wild corn over 80,000 years old. Columbus took some corn back with him to Europe, and today corn is a very important cultivated cereal plant in many parts of the world. The plant itself may be up to 6 m (20 ft.) high. The top of this annual contains the male ear and the leaf junctions contain the female flowers, which eventually form thin spadices (husks that will enclose the kernels).

Several corn kernels buried in regular potting soil (moistened) will grow quickly and form small plants. Put them in a pot or flat and keep the soil moist. A plastic bag, with air holes, wrapped around the pot or a lid placed over the flat will increase the humidity, which is beneficial. Transplanting may be necessary after a while. Regular watering, nourishment, light, and a sunny spot will insure that the plants do well indoors for a long time. Spray the leaves daily. If you want them to bear new kernels, place the corn plants in a sunny and protected spot in the garden. Several plants set close together will insure pollination; or hand pollination is possible.

If you would like to experiment and observe the seeds' growing power, place some kernels in a glass jar on a bed of moistened cotton, and put a lid on the jar. In a short time the sprouts will emerge, and you can then transfer the plant to a pot with soil. Using the same method you can also get wheat, rye, barley, and oats to grow. Fresh corn cobs that have been boiled or kernels that have not ripened will not grow.

Date Palm

Phoenix dactylifera

The date palm belongs to the *Palmae* family, and it is a feather palm whose wild origins are unknown. Today only the cultivated varieties are found. The date palm is one of the oldest cultivated plants, and in North Africa this palm has grown for thousands of years. It has been a very important plant for the inhabitants of that area. That the plant has had a great impact is demonstrated by Muhammed's words: "Honor the date palm that is your mother."

The date palm can attain a height of 15–20 m (50–65 ft.). The crown consists of about forty leaves. Each year about twelve of these die in succession and are replaced by new growth. The ancient Egyptians noticed this phenomenon and used the palm leaf to mark the months in their hieroglyphics.

Date palms are dioecious plants, which means that each tree has either female flowers or male flowers. Each cluster of flowers

will produce a great many fruits. In commerical cultivation, only a few male trees are planted in relation to the number of female trees. By various hand methods the growers assist nature in the pollination process. Even the ancient Babylonians knew that artificial pollination was necessary.

There is a saying about the date palm that it likes to stand with its head in the fire and its feet in the water. It tolerates much sun, but it must have access to ample water. The roots of the date palm can reach over 50 m (165 ft.) down into the soil. In its natural environment, the date palm is accustomed to temperature differences of up to 40–50°C. (72–122°F.) between day and night.

It takes a bit of patience to grow a date palm from a pit but it can be very rewarding. Wash the pits carefully in warm water, and let them dry before planting. It is fine to use regular potting soil or pure mulch. The planting container or pot should be rather deep, since the pit is drawn down into the soil by the root of the plant. This is a built-in protection that the seeds have, and it is extremely important in the desert, because this protects the seed from the heat.

Plant the seed about 3–4 cm (1½ in.) deep, and put the pot in a warm place. Spray the top soil daily. The soil should be kept moist, but it should not get too wet. If you tie a plastic bag, with air holes, over the pot, the humidity will be retained for quite a while, and watering need be done only at 10–14 day intervals.

It can take a long time before anything peeks up out of the pot, but eventually a tiny white shoot will come up—it looks like a toothpick. Several weeks later the shoot will turn green, and in time, you may see the plant grow into a palm tree.

Put the date plant in a light and sunny place, but remember that it can be easily damaged by too much heat from the sun if it is right next to the window. Keep the soil moist, and during the summer fertilize the plant. Spray the leaves daily with water at room temperature. During the winter, when a cooler location is advantageous, you can water less often, but do not forget the daily spraying of the leaves.

A more reliable way to get the pits to sprout is to lay them on a bed of moist cotton in a glass container, which is then put in a warm place. When the seeds begin to grow, which can take several months, they are planted 3–4 cm (1–1½ in.) deep in regular soil, in a tall pot.

Fig

Ficus carica

The fig belongs to the *Moraceae* family and grows up
to 6 m (20 ft.) high. It loses its leaves during the winter. The
fig is related to many of our common houseplants, and sometimes
you can find fig plants sold in the store. The wild fig probably
came from Asia Minor, although its origin is not definitely known.
The Latin name *carica* derives from the Kara landscape in
Turkey, near the city of Smyrna.

The figs most readily grown at home are the Adriatic, kadota,
mission, and brown turkey.

To grow a new plant from a fruit, pick out about ten seeds.
Plant these in a pot or flat and place it in a warm place. Use
regular potting soil or pure mulch. The soil should be kept moist,
but not too wet. Spraying the upper soil layer in addition to
covering the pot with a plastic bag, with air holes, permits
watering only once every two weeks.

When the shoots come up, let the largest one keep growing,
and snip off the lesser ones. This makes it easier for the remaining
shoot to grow. After a while, the plant should be transplanted and
eventually placed in a fairly large pot with regular soil. The soil
can be rather heavy and should contain some clay.

During the summer, the fig should stand in a light and sunny
place, with protection against the strongest midday sun. Water

regularly and generously, but remember that the plant should dry out somewhat between waterings so that the roots can breathe. Rich fertilizer is needed during the summer months. In the autumn the plant loses its leaves, and then it should be put in a cool place. Be sparing with water during the winter, when the plant is dormant.

A plant can form fruit indoors, especially if it has the space to grow properly.

A major commercial variety of fig is the Smyrna fig, or as it is called in some areas, the Calimyrna fig. However, since all the flowers of this type are female, the Goat fig (*capricus*), which has female and male blossoms, is cultivated along with it. The blossoms are situated on the inner side of the flower joint and are never seen from the outside. The flower joint eventually becomes the skin of the fruit.

Pollination is taken care of by insects, especially by the fig wasp, which lays its eggs in the Goat fig's blossoms. When the eggs open, the wasps fly out, and the females, which are covered with pollen, fly from one Smyrna fig blossom to another. In this manner, they fertilize the Smyrna fig. The females do not lay eggs in the Smyrna fig, but wait until they find a new Goat fig tree.

This method of pollination, called "caprification," was known for thousands of years, and to aid the process people used to throw branches from the Goat fig up into the Smyrna fig's crown. However, the growers did not do this when they first tried to raise figs in California, and the endeavor was a failure until they planted the Goat fig and imported fig wasps.

There are, of course, some figs that can mature and bear fruit without this process.

After pollination, the fig blossoms that sit in the blossom envelope become enlarged, and a multiple-skin fruit is formed. Fresh figs usually have white or bluish-brown skin and are sold quite often in the stores, especially in early and late summer, at least in the larger cities.

Grapefruit

Citrus paradisi

The grapefruit is a member of the *Rutaceae* family. Discovered by Europeans for the first time in Puerto Rico in 1759, the fruit was later renamed grapefruit. This unusual name is thought to have something to do with the way that both grapes and grapefruits grow, in clusters.

This citrus fruit is now cultivated in favorable climates all over the world.

For the planting of grapefruit seeds, see Orange, page 109.

Grapes

Vitis vinifera

The grape is a member of the *Vitaceae* family, which also includes many common houseplants. Grapes were an anciently cultivated plant, and they had acquired great significance long before the Egyptian empire was established on the Nile. There are a number of wild types in the world, including several edible types in America. The cultivated types are thought to have had their origin south of the Caspian Sea, in what was called Armenia. About 1,500 different kinds of grapes are now grown. Many are important in wine production—others are cultivated for their edible fruit. Some types are grown for both purposes.

The seeds of grapes are easy to plant, using regular potting soil that is kept only slightly moist. Put the pot in a warm and light place. When the seedling comes up, the soil is watered intermittently so that air can penetrate down into the roots. Generous fertilizing will contribute to success. Spray the leaves occasionally with lukewarm water. Flowers can develop, but not grapes on a plant kept indoors. In many areas, you can put the plant in a protected place and grow it in the garden. At the end of a growing season the plant should be pruned to encourage the production of more fruit the following season.

Guava

Psidium guajava

The guava belongs to the *Myrtaceae* family, which also includes myrtle and eucalyptus trees. The plant is a native of the West Indies and Central America, but it is now grown in most subtropical regions.

The guava is a very sweet-tasting fruit, and when cut open it resembles a melon, with its seeds grouped near the center. The plant itself has beautiful leaves with a clear-veined pattern and a downy underside. The flowers are large and white, and the fruit is golden or yellow-green. Guavas are often used to make jam.

Guavas can be purchased during the fall. Pick out a few of the seeds, wash them carefully in warm water, and let them dry for several days. Then sow them in a pot or flat in regular potting soil. The soil should be well-moistened when planting. A plastic bag, with air holes, or a glass cover or transparent lid over the flat increases the humidity, and then you need only water at long intervals. Put the pot in a warm location.

As the plant grows it will be necessary to transplant it into a larger pot. Prune it to induce branching.

Light, warmth, and sun, with protection from the midday rays, are the requirements for a healthy plant. During the summer it is recommended that you water regularly and generously fertilize the guava. During the winter water it somewhat less. Regularly spray the leaves during the entire year. Under good conditions, flowers may blossom in several years. These can be artifically pollinated and afterwards the fruit will develop.

Hazelnut

Corylus avellana

The hazelnut belongs to the *Corylaceae* family. It grows wild in some parts of the world, and is cultivated commercially in many other areas.

Guava

Hazelnut

The hazelnut actually is produced by a bush that can reach up to 9 m (30 ft.) tall, but in northern latitudes it seldom reaches over 5 m (17 ft.) .

For hazelnuts to sprout, they must be well-wintered. This is a built-in protection for the nuts, so that they will not grow in the autumn and then freeze during the winter. In cold climates lay the nuts in a pot with lightly moistened sand or moss, and then place the pot on a balcony or in the garden. They should stay there about three months. An ideal temperature is 1–6°C. (34–45°F.) . If they are kept out in the cold slightly longer there will be no harmful effect, but if they are to be kept out a much longer time first insulate the pot with newspaper or some other material. For "wintering" in warmer climates, see page 44.

When you are ready to plant the nuts, sandpaper or file down the shells so that the sprout has an easier time breaking out. Use regular potting soil, and preferably a deep pot with plenty of room for the developing roots. Bury the nut a few cm (1–1½ in.) , and keep the soil moist but not too wet. Spray the soil's surface every day with water that is at room temperature. Put the pot where it will have sufficient light.

After several weeks, sometimes longer, the sprout will come up. The pot should be moved to a relatively cool place, where the humidity is high; otherwise damage will result in the form of brown spots on the leaves. Daily spraying, regular watering, and fertilizing will provide favorable conditions for the hazel's growth.

Do not keep the hazel as a houseplant for a long period of time. Plant it outside in the garden or in the wild to provide some of the other conditions necessary for its survival. Because the hazelnut plant is dioecious, or with both male and female types, it is necessary to have both kinds in order to produce nuts.

Kiwi

Actinidia chinesis

The kiwi belongs to the *Actinidaceae* family and it is native to the Yangtze River valley in China. It has tendrils and grows by clinging to other plants.

The kiwi is cultivated similarly to the way grapes are cultivated—arbors are built, and on these the kiwi's tendrils can attach themselves. The fruit hangs like grapes, except that instead of clusters of fruit, there is just one fruit.

Most kiwis come from New Zealand, and they are often flown from there. Kiwis can also be grown in the United States, and they ripen in the summer.

The kiwi is actually a berry. It is beautiful when cut open, with the almost black seeds standing out strikingly against the green flesh. The skin is downy and brown.

To be successful in cultivating the kiwi at home, the seeds

should first be wintered. Remove several of the seeds very carefully with a knife and spread them out to dry for several days on some newspaper or brown paper. Then lay the seeds in lightly moistened sand or moss in a pot or other type of container, and put it in a dark spot, at a temperature of 1–6°C. (34–45°F.) . The best place for this is the refrigerator. If you put the seeds in a tightly closed jar, this is not at all unhygienic even if you place the seeds near food.

When wintering the seeds, it is not too important for the sand or moss you use to be sterile; but when sowing them, sterility of the soil is essential for effective results.

Sow the seeds in a pot or flat and use a very light and porous soil. Packaged potting soil is ideal, because it is sterile. The soil should be very moist. Spread the seeds out in a pot and cover them with a thin layer of soil. Spray the upper soil layer every day with warm water so that the humidity is kept high, and then you will not need to water as often. Place the pot or flat in a warm and light place, where it is sunny.

After about eight weeks, usually, the plants will come up. The leaves are heart-shaped with silvery down. Water the plants regularly, but remember that surplus water will cause them to wither. Put the pot in a light place, with protection against the strongest midday sun. Fertilizer should be applied during the warm seasons.

During winter, the kiwi plant may lose its leaves, but new leaves will appear in the spring. You should arrange some sort of support to which the kiwi can cling as it grows.

It can't be said for certain whether blossoms and fruit will develop indoors. In any case, this plant, too, is dioecious, requiring both male and female types for fruit production.

Kumquat

Citrus fortunella

The kumquat belongs to the *Rutaceae* family and came originally from China. Kumquats are a very common fruit in China and Japan, and formerly they were called *C. japonica*. During the mid-nineteenth century it was usually cultivated only as a decorative plant. Now it is cultivated as a fruit in Northern Africa, California, and Florida, among other places, and the fruit can be found on the market during the winter.

The kumquat is small, oval, and orange in color, but sometimes round fruits are produced. They are often sold still attached to their leaves or with one or more fruits on a tiny branch. The fruit is eaten with the skin.

For the planting of kumquat seeds, see Orange, page 109.

Lemon

Citrus limon

The lemon belongs to the *Rutaceae* family and originated in the region south of the Himalayas. The fruit was known in China long before recorded time, and it came to Europe around 100 A.D. This is not absolutely certain, because the lemon tree is often confused with another variety of citrus that is actually an entirely different plant. However, we do know that lemons were grown in Spain and Sicily in the thirteenth century, and that this fruit came to Europe several hundred years earlier than its relative, the orange.

The fruit contains a high percentage of lemon juice and vitamin C, which has long been known to be helpful against the terrible disease, scurvy. It was written in 1795 that lemons should be included in the supplies on board the British Navy's ships.

The flowers of the lemon are usually white, but sometimes they are pink. The leaves are green, and can be distinguished from the orange by the lemony aroma of the crushed leaves.

For the planting of lemon seeds, see Orange, page 109.

Lime

Citrus aurantifolia

The lime belongs to the *Rutaceae* family and probably came from East Asia, although we are not quite sure of its origin. It was carried to all parts of the world, including the West Indies and Latin America, where it is now quite common. The tree is between 2–4 m (6½–13 ft.) tall, and resembles a lemon tree. The skin and pulp are green, and the shape of the fruit is like that of the lemon. The skin of the lime is very thin.

For the planting of lime seeds, see Orange, page 109.

Litchi Nut

Litchi chinensis

The litchi nut is over 4,000 years old. It is an ancient Chinese cultivated plant belonging to the *Sapindaceae* family. In China, the litchi is planted near rivers and canals and at the edge of rice fields—in the same way that willow trees are planted in some areas.

The fruit that is sold often has a brown, rough shell; and, inside, the pulp is white and gelatin-like, with a quite large, brown pit.

Unfortunately, this seed has lost viability, and it is not worthwhile trying to plant it. However, fresh fruits have a red or reddish-brown skin, and if you plant them within five days of the harvest, at the latest, you can succeed in growing your own litchi nut tree at home in your window.

In many northern areas you cannot get the fresh fruit, but in some warm climates, you may be able to find fresh litchis. The fresh fruit is very good, and it is easy to get the seeds to grow.

You should plant the seed immediately after you have eaten the fruit. The seed should buried horizontally under moist soil, about 2 cm (1 in.) deep. It is fine to use regular potting soil, and in a wide pot you can plant several seeds. Keep the soil moist. Since it is important that the soil be able to breathe, you should put several cm (1–1½ in.) of gravel or coarse sand in the bottom of the pot.

Put the pot in a light and warm place, but when the plant comes up, it must be protected from strong sun. The plant is very sensitive to drafts and cold, like many other subtropical plants. Water regularly, but never let the water stand in the outer pot or plate. Spray the leaves regularly with warm water, using a fine spray. Fertilize during the summer months.

Another good method of getting litchi nuts to grow is to place the seed in a glass jar, on a slightly moist bed of cotton. When the sprout comes up, plant it in soil.

There will be neither flowers nor fruit on a plant kept indoors. In their natural environment, litchis take about ten years before the first blossoms appear. The tree can live to be quite old, and there are examples today of trees over 800 years old.

Loquat

Eriobotrya japonica

The Japanese loquat is native to China but is now found in many parts of the world. It has large green leaves with strongly marked veins and a downy brown underside. The fruit is golden and plum-like, and inside the fruit there are several large seeds.

The loquat is quite perishable and that is why it is not sold in many northern regions. If that is where you live and if you should travel south in midwinter or spring, you will likely encounter this fruit, and its seeds can then be planted at home.

The seeds are planted in regular potting soil, in a roomy pot. Keep the soil moist and place the pot where it is warm. When the plant comes up, it should be moved to a place with light and provision for shade to protect against strong sun. Regular watering and fertilizing should be done during the warmer seasons. Be somewhat careful in the winter about watering, and put the pot in a cooler location.

The loquat as a houseplant can look truly beautiful in your window. The blossoms usually appear in November and December, and the fruit ripens in the spring. However, blossoms or fruit may or may not develop inside the house.

Mango

Mangifera indica

The mango belongs to the *Anacardiaceae* family and comes originally from India and Southeast Asia. In the northeastern part of India, the mango has been cultivated for more than 4,000 years. It is one of the most common large fruits cultivated in the tropics. Since the seeds of the mango are viable for a very short period of time, the tree did not spread to other parts of the world until faster ships were developed during the nineteenth century.

There are many different types and hybrid varieties of mango, having a wide range of shapes and colors. Yellow and orange are the most common colors of the fruit. The seed found in the mango can be planted, and it takes between three weeks and four months before the seed will sprout.

Remove the seed carefully and wash it in warm water, so that all of the pulp within the fibers around the seed is removed. The fibers themselves should remain. Since the seed can cause allergic reactions, allergy-prone people ought to use protective gloves.

After washing, the seed should soak in warm water in a warm place for about five days. Change the water every day. Then plant the seed with the eye upwards, vertically, in very light soil. Use packaged potting soil and plant in a large pot. Bury the seed about one cm (½ in.) under the soil.

The soil should be well-moistened when planting the seed. Afterwards it should be watered relatively often, compared to other seeds and pits, but avoid water-logging the soil. Always use warm water and spray the top soil daily. A plastic bag, with air holes, or a transparent lid over the pot increases the humidity, which will promote growth.

When the plant comes up, put the pot in a light and warm place. Mangos are very sensitive to the cold. Putting the plant too close to the windowsill during winter can cause the flowers to wither. Regular watering is recommended for the entire year, and fertilizers should be used during the warmer seasons. Transplanting may be necessary after a while, because the mango does best in a relatively large pot. Spray the leaves daily.

During the autumn, you should set the older plants out for a short dry spell, which the plant is used to in its natural habitat. In this way, the flowers will be more prone to develop later on. During this period, water the plant only half as much as usual.

After a minimum of four years the tree is able to produce flowers. These blossoms ought to be artificially pollinated, and they can, with luck, result in fruit.

Melon

Cucumis melo

Melons belong to the *Cucurbitaceae* family, along with the cucumber, squash, and pumpkin. The common melon, or cantaloupe, is of Indian origin and was cultivated by the ancient Egyptians. Watermelon, *Citrullus vulgaris,* came from Africa, where it is still found growing wild. Watermelon was cultivated for thousands of years in Egypt and India.

The seeds of the melon should be thoroughly washed and allowed to dry for several weeks in a dark place, in an airtight jar or plastic bag, before they are planted. Use regular potting soil in a roomy container. Bury the seeds 1–2 cm (½–1 in.) deep, in well-moistened soil. A glass cover or a plastic bag, with air holes, over the pot will increase the humidity. Spray the soil's surface daily, and water the plant periodically so that air can reach down into the soil.

Melon seeds are fast sprouters, and the plant develops quickly. In order for blossoms and fruit to form, the melon plant should be set out in the garden or on a porch or balcony.

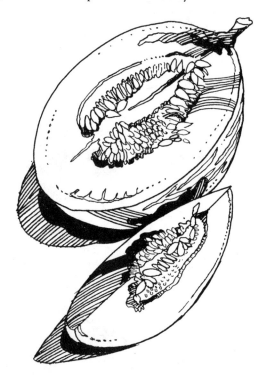

Nectarine

Prunus persica, var. nectarina

Nectarines belong to the *Rosaceae* family and are a type of peach with a smooth skin. They can be found in stores during the summer.

For the planting of nectarine pits, see Apricot, page 55.

Olives

Olea europaea

The olive tree belongs to the *Oleaceae* family (to which lilacs and forsythia also belong), and it evolved from a wild tree in the Orient. The olive is an anciently cultivated plant, described and discussed many times in the Bible. The dove that Noah released from the ark to see if the water had subsided came back with an olive branch in its beak.

The olive tree grows very, very slowly and may live to be up to 2,000 years old. The tree lends a characteristic appearance to the landscape, with its grey-green leaves and often gnarled trunk. The leaves are spear-shaped and downy grey-green on the underside. They resemble silver arrow leaves, and they always grow out from the branches in pairs. The leaves are able to turn their undersides towards the sun, and thus protect themselves during midday heat or drought.

In the junctures of the leaves bloom whitish-yellow flowers, with a slight fragrance. The fruits, which are green at first, change to a blue-black color with maturity. The pulp and pit contain valuable olive oil. The olive is a stone fruit.

The pits from ripe olives can be planted. Sometimes fresh olives are sold in the stores, at least in the larger cities. If you live—or have friends who live—in warm climates, getting untreated olives should not be too difficult.

For the planting of olive pits, see Apricot, page 55.

Onion

Allium cepa

Onions came originally from West Africa, but they no longer grow wild there. There are many types of onions that we use in cooking, and they all belong to the *Liliaceae* family, to which many of our common houseplants and cut flowers also belong.

Growing onions indoors is a delightful experiment that everyone, including the children, can enjoy. It will interest and surprise you to discover the growing power of an onion.

Put the onion in a glass, preferably a French hyacinth glass. Fill it up with water until it just reaches the onion. After a while roots will emerge and a shoot will grow up out of the top. If you allow the onion to grow for a long enough time, a beautiful flower will bloom.

Even garlic, *A. sativum,* can be grown in a glass or planted in regular potting soil. The taste of the shoot reminds one of a green onion, and it can be clipped and used to spice up sandwiches or salads.

Leeks, *A. porrum,* can be planted in a tall vase with a little water in the bottom, or in slightly moistened sand.

Chives are also easy to grow and delicious to eat.

Orange

Citrus sinensis

Sweet oranges are members of the *Rutaceae* family and originally came from Vietnam and China, where they have been grown for thousands of years. The first entry of oranges into Europe occurred in the fifteenth century, and later they were carried further by the Spaniards, to the Americas. Now oranges are grown in many parts of the world and over the years a number of varieties have been developed.

Commercially-grown trees can each bear several hundred fruits at the same time. The fruit is actually a berry with a pulp of juicy threads, and over 70 percent of all oranges are used to make juice. In commercial production, cultivation occurs by grafting.

Raising orange trees from seeds is quite simple, and if the plant is topped and well cared for, it can bloom in 3–5 years. Dwarf varieties are especially good for pot culture.

The oranges that are sold in the stores usually have a number of seeds, and generally these are viable. Wash the seeds in lukewarm water and let them dry before planting. Oranges grow well in many soil types. Common potting soil, possibly with a mixture of thinner flower bed soil and 10 percent sand, is ideal for you to use in planting orange seeds. Plant several in a pot or flat.

Place the seeds beneath the soil's surface, and give them a good watering. Then let the soil dry thoroughly before the next watering. If you cover the pot with a plastic bag with breathing holes, it should take about 10–14 days before you will need to water again. Spray the top soil with warm water daily, so that the humidity level will be kept high.

Place the pot in a light location with warmth nearby, such as near a radiator. It is also advisable to provide extra light, but remember that then you will need to water more often. How long it takes before the seeds will grow is dependent, among other things, upon the season.

When the plant is 10 cm (4 in.) high, plant it in its own pot. Additional transplanting may be necessary later on, since orange trees thrive in relatively large pots. After a while, it will also be necessary to top the plant in order to develop well-branched, bushy plants.

Place the orange tree in a light and sunny place. In the summer place it by an open window or outdoors. Remember that direct sunlight can be very hot, and a sun shade for midday may be necessary. In the winter, a cool (10–15° C. or 50–59°F.) location is fine. Watering should be periodic, so that air can circulate down to the roots. Fertilize the plant during the warmer seasons.

Spray the leaves often with warm water and regularly check the plant for pests. Red spider mites and aphids have a special love for citrus. (See Pests and Plant Diseases,page 45.)

The sweet-smelling flowers, white with lighter purple trim, may blossom in three years, but it often takes longer. For fruit to develop, the flowers must be pollinated. This is done simply by transferring pollen with a brush or feather from flower to flower. Oranges require 10–16 months before they will develop and ripen.

Citrus plants are decorative, even without fruit, and can be a beautiful addition to your home environment. Older plants often have thorns, as well as flowers.

Clementine Orange

Citrus reticulata var. clementin

The clementine (or Algerian) orange belongs to the *Rutaceae* family. It is the product of a cross between mandarins and Seville sour oranges, achieved by an Algerian priest named Clement in 1902. The fruit from younger plants contain few seeds, while the fruit from older plants can have many seeds. The clementine is rather flat in shape, as are most of the small citrus fruits.

For the planting of clementine seeds, see Orange, page 109.

Mandarin Orange

Citrus reticulata

The mandarin orange belongs to the *Rutaceae* family and originally came from the Philippines. The name derives from the island Mauritius (Mandara), east of Madagascar. Mandarins have a large number of seeds.

For the planting of mandarin seeds, see Orange, page 109.

Satsuma Orange

Citrus reticulata var. unshiu

The Satsuma orange belongs to the *Rutaceae* family and is native to Satsuma province on the Kyushu Island of Japan. The Satsuma is a relative of the mandarin, and can be purchased from fall through spring. The Satsuma can be confused with the clementine orange, but the Satsuma is a much less expensive fruit than the clementine. Tangerines are another type of mandarin.

For the planting of Satsuma orange seeds, see Orange, page 109.

Seville (Sour) Orange

Citrus aurantium amara

The Seville or sour orange belongs to the Rutaceae family and came originally from Indochina.   In the ninth century the tree reached Arabia, and early in the twelfth century the first trees were planted in Europe, in Sicily. The skin is very aromatic and is used in marmalades and in baking. The pulp is very sour.

Bergamot, C. aurantium bergamia, is cultivated primarily for the golden bergamot oil found in the skin. This is used in various perfumes.

For the planting of Seville orange seeds, see Orange, page 109.

Papaya

Carica papaya

The papaya belongs to the *Caricaceae* family. It originated in the West Indies and Latin America, where it resulted from the crossing of wild types. The papaya is a tropical tree that may reach up to 6–9 m (20–30 ft.) tall, and the leaves are large and deeply scallopped. The fruit is pear-shaped with yellow-orange skin. The pulp is also a yellowish-orange and butterlike in consistency, with hundreds of seeds concentrated in a hollow space in the center of the fruit. The papaya can be found in many stores throughout most of the year.

To plant a papaya, scoop out the seeds with a spoon and spread them on a newspaper to dry. Each seed is surrounded by a hard shell and if you press down on it, the seed itself will pop out. Remove this shell before sowing the seeds, otherwise they will take a very long time to sprout. The inner seed should then lie on a newspaper to dry for about 24 hours before planting.

Use very porous soil for the papaya and a roomy clay pot in which you can sow 10–12 seeds. Put gravel or coarse sand several cm (1–1½ in.) deep in the bottom of the pot, and fill the pot with well-moistened soil. Cover the seeds with about one cm (½ in.) of soil. The pot should be placed in a light and warm location, and since it is very important for the soil to get enough air, water it only intermittently. It does not hurt to spray the soil surface with warm water daily. If you put a plastic bag, with air holes, over the pot, be sure the holes provide adequate ventilation for the soil.

In a short while , the seedlings will come up. Thin out the least developed ones in the pot, so that eventually only one plant remains. If you pull out these smaller seedlings carefully, retaining the soil around their roots, you can try to plant them in another pot—although the papaya is a very sensitive plant.

The seedling likes to be in the light, and the plant's size will vary depending upon the amount of sun it gets. Spray it daily with warm water. Regular watering at intervals so that air can penetrate into the soil, as well as fertilizing, will aid growth during the warmer seasons.

The papaya is very sensitive to water that remains standing in the pot or dish. It is also sensitive to air pollution, so you might not have luck in growing papaya plants if you live on a street with heavy traffic. Small plants are also easily infected with certain diseases, but if you do have success, you can obtain a very beautiful houseplant that can live for many years.

The papaya contains papain, a proteolytic (protein-splitting) enzyme, which is used as a meat tenderizer, in toothpaste, and in cosmetics.

114

Passion Fruit

Passiflora

Passionflower plants belong to the *Passifloraceae* family and most of the 400 known types come from tropical Latin America. The fruit is called granandilla, a name given by the Spanish explorers who thought that the fruit looked like a small pomegranate, which is what granandilla means. The passionflower is a climbing plant or bush. The blossoms on a number of varieties are very large and magnificent, and these types are among our common houseplants.

A Jesuit monk on an assignment in Peru is said to have had a revelation when he saw the passionflower. He thought that he could see before him the story of the passion of Jesus in the flower's strange structure, and this is how the plant acquired its name.

The passionflower is a climbing plant or bush. The blossoms on a number of varieties are very large and magnificent, and these types are among our common houseplants.

Your friends will surely marvel at this remarkable plant which will quickly develop a green and then later yellow fruit the size of a plum. This fruit, or actually berry, is delicious to eat. Inside the berry there are many seeds which can be used to reproduce new plants. Passionfruit is available in some stores in the late summer and fall. The most important varieties are *P. edulis* and *P. laurifolia*.

Dig the seeds out with a spoon and wash them thoroughly in warm water. Let them dry for 24 hours. Plant them in regular potting soil in a flat or a pot. The soil should be moistened when planting, and the seeds should be buried about one cm (½ in.) under the topsoil. You can also plant the seeds in pressed peat pots (Jiffy pots) that swell up when watered, and transplant them into larger pots when the plants have grown and need more room. Put the flat or pot in a warm place and keep the soil moist. Spray the topsoil with warm water every day, and put a plastic bag, with air holes, or a glass lid over the flat to help retain the humidity.

When the seedling comes up, put it in a light and sunny place with protection from the strongest midday sun. Water regularly and give it fertilizer during the warmer seasons. Spray the leaves daily with warm water. Transplanting may be necessary when the plant needs more space. Artificially pollinate the flowers by transferring the pollen from one blossom to another with the help of a brush or feather. The fruits will ripen in the late summer and autumn.

In the winter, the leaves of the passionflower fall off, and the plant should then be watered less, and not given any fertilizer. Early in the spring, prune and transplant.

The passionflower is a tropical plant and sensitive to the cold and to drafts.

Peach

Prunus persica

The peach belongs to the *Rosaceae* family. It originated in China and is a close relative of the apricot. Soldiers of Alexander the Great came in contact with this tree in Persia during that great eastern campaign, and this is how the peach acquired its name.

The peach is a rather short tree with narrow, spear-shaped leaves and very pretty pink flowers that blossom early in the spring on the tree's bare branches. The fruit is downy and yellow, with a reddish blush on the sides exposed to the sun. The pit is rough and is easily grown into a new plant. In many areas, the peach can grow outside the entire year in a protected place.

For the planting of peach pits, see Apricots, page 55.

Peanut

Arachis hypogaea

Peanuts belong to the *Leguminosae* family, along with beans and peas. They came originally from South America. The peanut is a beautiful plant with golden sweet pea-like flowers, which bend towards the earth after pollination and dig themselves beneath the surface. Subsequently, the pod develops underground, with the nuts growing inside the pod. The plant lives one year.

Unroasted peanuts still in their shell or pod should be purchased, and the two seeds that are found in each shell can be sown. After taking the nuts out of their shell, plant them in

119

regular potting soil in a pot or flat. Pressed peat pots (Jiffy pots) that swell up when watered are good to use for sowing peanuts.

The soil should be kept moist, but not too wet.
A plastic bag, with air holes, over a pot or a glass cover or a transparent lid over a flat will increase the humidity. Spray the surface soil daily, and water at 10–14 day intervals. Check in between waterings to make sure that the soil is not too wet. Put the pot in a warm and light place, preferably near a source of heat.

When the plant has grown somewhat, it should be transplanted; a wide pot will enable the plant to continue growing. Put the pot in a spot with warmth and light, with shade protecting against the strongest midday sun. Regular watering and fertilizing according to directions are required. Use the same fertilizer that you use for other houseplants. Spray the leaves in the morning and evening with water at room temperature.

When the flowers blossom, carefully pollinate them by

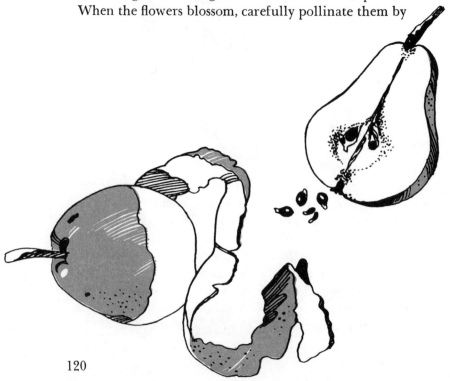

carrying the pollen from one flower to another with the help of a feather or brush. With luck, the nuts will then develop.

Pear

Pyrus communis

The pear belongs to the *Rosaceae* family, and wild types through breeding have given rise to the cultivated pears from Asia, Europe, and North America.

The pear tree is pyramid-shaped, as distinguished from the apple tree which has a round shape. The flowers form in clusters and are white in color. There are over a thousand varieties of pear.

For the planting of pear seeds, see Apple, page 53.

Peas

Pisum sativum

Garden peas belong to the *Leguminosae* family, and have evolved from the wild varieties called *P. ealtius*. There are many different types of peas, and most of them are interesting to experiment with to see what growing power these seeds have.

Line the inside wall of an ordinary drinking glass with thick paper, preferably blotting or absorbent paper. Fill the resulting paper "tube" with soil or sand that has been well-moistened. In the tight space between the blotting paper and the glass insert the peas. They will absorb moisture from the soil or mud—through the paper—and you will be able to observe their development.

You can also put a couple of yellow peas in moist plaster, and watch how the peas quickly spring out of the hardened plaster.

Planting peas is really something of an experiment, and you cannot expect to keep the plants indoors for a long time. You can also plant peas in regular soil in a pot and then see how long the plants can survive inside the house. Peas are cared for in the same manner as beans.

121

Pecan

Carya illinoensis

The pecan tree belongs to the *Juglandaceae* family and comes from the southern regions of the United States. It belongs to the same family as the walnut, but the fruits or nuts of the pecan have a brown or reddish-brown, smooth shell.

Sandpaper or file the nut so that the sprout will have an easier time breaking through the shell. Use a deep pot. Add well-moistened, regular potting soil on top of about 2 cm (1 in.) of gravel in the bottom of the pot, which will provide adequate ventilation of the soil. Bury the nut about 3 cm (1¼ in.) deep, and put the pot in a warm and light location.

Place a plastic bag, with air holes, over the pot to retain the humidity. Spraying the upper soil layer with warm water will mean that you need only water at long intervals.

When the shoot comes up, move the pot to a sunny location. Spray the leaves daily and keep the earth moist, applying a generous amount of fertilizer during the summer. It can be difficult to get the pecan tree to survive for a long time indoors, but it is a beautiful tree as long as it lasts.

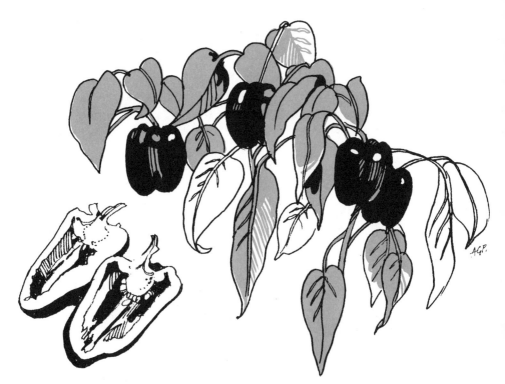

Pepper

Capsicum annuum

The pepper family consists of more than 2,000 varieties—the fresh peppers that we eat raw or baked, the Spanish pepper or pimiento, and others, including small types such as those ground into cayenne pepper. They all belong to the *Solanaceae* family.

Peppers came from South America. They were brought to Europe in the sixteenth century and then quickly spread to the warmer parts of Asia. Today's peppers evolved from the small fruiting types. These small types—notably paprika—have been developed extensively in Hungary.

The blossoms of the pepper plant are small and white, and the fruits vary in shape. Unripe pepper's unique taste comes from a chemical called capsaicin, which is found in heavy concentration in the fruit.

The seeds can be planted if they are first spread out and dried in a warm, light place for several days. Sow several seeds of one type of pepper in a flat or pot. Put them in a light and preferably warm location. Cover the pot with a glass or a plastic bag with air holes, so that humidity is retained. Keep the soil moist.

When the plants come up and have grown about 3 cm (1½ in.), transfer them to pressed peat pots (Jiffy pots) that swell up when watered. You can also sow the seeds in peat pots at the beginning. When the pepper plants have grown to about 10 cm (4 in.), put them into pots with regular potting soil. Keep them in

a warm, light, sunny place with protection from the strongest summer sun.

It is fine to give the small pepper plants extra lighting in the winter. The pepper plants can be put outside in a protected and sunny spot during the summer. Accustom the plants to the outdoors gradually, until eventually they can stay outside the entire day.

Spanish Pepper (Pimiento)

Capsicum annuum

The Spanish pepper or pimiento belongs to the *Solanaceae* family and is similar to the common pepper plant, but it is not as popular nor as widely cultivated. Spanish pepper as a houseplant will shine beautifully in the window, with red, yellow, or purple colored fruits on the branches. The blossoms are small and white. The fruits can be used in cooking, but they are very spicy, so one should be careful when using them. The seeds of the Spanish pepper can reproduce new plants.

For the planting of Spanish pepper seeds, see Pepper, page 123.

Persimmon

Diospyrus kaki

Persimmons belong to the *Ebenaceae* family and originated in China, Japan, and Korea. They grow on a tree that can reach 2–4 m (6½–13 ft.) tall, with spirally set leaves that are oblong and shiny green. The blossoms are small and reddish-white. The fruit is the color of an apricot, golden orange, while the shape of the fruit and its skin resemble that of a tomato. The persimmon tastes somewhat like a peach.

Persimmons are sold in the stores in late fall. Inside the fruit there may be as many as eight black seeds, and these can be sown and will reproduce new plants.

The persimmon seeds should be wintered first (see page 44).
They are then planted in moistened potting soil in a pot or flat,
about one cm (½ in.) deep. Put the pot in a warm and light
place, and keep the soil moist. It is important to give air a chance
to penetrate into the soil between waterings. Daily spraying of the
upper soil surface, as well as putting a plastic bag, with air holes,
or a lid over the pot or flat will increase the humidity.

When the shoot comes up, it should be put in a light and
sunny spot with protection from the strongest midday sun. Water
regularly with generous amounts of fertilizer during the warmer
seasons. When the persimmon grows larger, transplant it into a

bigger pot. It does well in relatively light soil with some clay. To increase ventilation of the soil, put sand or gravel in the bottom of the pot.

The persimmon likes warmth and moisture, so daily spraying of the leaves is a must. Neither flowers nor fruit will develop indoors, but the persimmon can become a beautiful houseplant.

Pineapple

Ananas comosus

The pineapple is a member of the same family that some of our common houseplants belong to—the *Bromeliaceae* family. This name derives from the surname of a Swedish doctor and botanist, Olof Bromelius, who lived in the seventeenth century. Practically all of the plants in this family originated in the tropical regions of South America; pineapples, for example, came from Brazil and Paraguay.

Long before the Europeans came to the Americas, the Indians cultivated pineapples in large areas of South America. On Columbus' second voyage, he saw the inhabitants of Guadalupe Island eating this fruit. Since he thought that it resembled a large pinecone, he gave it the name pineapple.

In a very short period of time, pineapples were spread by the Spaniards and Portuguese to the warm regions of the world, and large plantations were started. In the beginning of the nineteenth century the fruit was introduced to Hawaii, and there today is the largest commercial cultivation of pineapples.

Pineapples, like strawberries, are classified as a complex accessary fruit, because of the way they grow. The pineapple spike has more than one hundred flowers which emerge from the leaf rosettes (a rosette is a cluster of leaves that radiate from a central stem). The flowers become enlarged after fertilization, and they fuse together with the main leaves and stalk to create a fruit.

To grow a plant from a pineapple that you buy in the store, make two diagonal incisions under the leaf rosette (or crown), cutting down into the fruit. Take about 2 cm (1 in.) of the fruit as you cut off the rosette. Be sure that you get an undamaged rosette.

This rosette now has to sit about two days to dry, and then

you can plant it in regular potting soil. The soil should reach up to the beginning of the outer leaves. Use a clay pot, and later transplant the pineapple into a larger pot. Some plants of this family like relatively small pots, however.

Keep the soil moist, but make sure that it is not too wet. Put the pot in a light and warm place, preferably setting it in a plastic bag, with air holes. In this way the humidity will be kept high, which the pineapple likes. You should also water the leaves with a vitamin or fertilizer solution, using a 10 percent concentration. Pineapples have poorly developed roots, but they can absorb water and nourishment through their leaves.

After 6–8 weeks, when the roots have developed, the plastic bag can be removed. The plant should continue to be located in a light but not sunny spot, and watering should occur only at intervals so that air can penetrate down into the pot. It is good to spray the leaves often, using a fine spray. Add fertilizers to the soil, but be careful. Too much fertilizer will result in leaf damage, in the form of brown spots or points on the leaves.

Sometimes a pineapple plant will grow a side shoot, and if you plant this, you can produce a new plant. Before you separate it from the mother plant, the side shoot should have at least six well-developed leaves. Detach it carefully from the mother plant, and plant it in a pot the same way you did the original rosette. In commercial pineapple cultivation, it takes 18–30 months for a plant to bear fruit. Indoors, you should count on it taking somewhat longer, if it produces fruit at all.

To obtain good results with pineapples, put a pinch of iron powder (available at pharmacies or from chemical supply stores) into the soil at planting time, and again when the plant is ready to bloom. Setting the plant near a source of ethylene gas will also have a beneficial effect. Ripening apples give off a lot of ethylene gas, and a bowl of this fruit in the vicinity of your pineapple plant can stimulate it to bloom.

When blossoming begins, the plant should be moved to a sunny spot. Otherwise, the fruit will not develop.

Plum

Prunus domestica

The plum belongs to the *Rosaceae* family, and the cultivated varieties are crosses between the Asian cherry plum and the common wild plum or sloe. There are a number of varieties of plums, and the fruits are found in many colors. The original plum type, *bullace,* can also be crossed with the cultivated varieties. The plum was carried to Europe from the East during the Middle Ages.

Sometimes plum pits are sterile, and you cannot tell this by looking at them. This is why, unfortunately, they often do not grow when planted. Still, there is a fair chance that you will be able to grow a little plum tree in your window.

Since the plum is not always true to its type when raised from a seed, you never know what kind of plum you can expect to get. In commercial cultivation, plum trees are reproduced by grafting.

The plum pits must be wintered before planting. For further instructions, see Cherry, page 71.

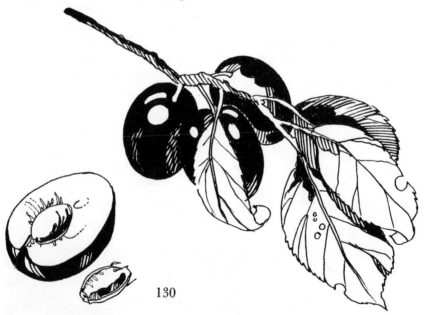

130

Pomegranate

Punica granatum

The pomegranate belongs to the *Punicaceae* family, which is a near relative of the myrtle. The tree probably came from the Middle East and Mediterranean regions and it has been cultivated for a very long time. It is deciduous and hardy. It makes a delightful bonsai, or miniature, tree. The fruit is a commonly used motif in various art forms and was often incorporated as a pattern in tapestries during the Renaissance.

The pomegranate was regarded as a holy plant both by the Egyptians and the Jews. The Greeks used the fruit as a symbol of fertility. The Spanish coat of arms also uses the symbol of the pomegranate. The Latin name for pomegranate, *Punica,* was given by the Romans who imported the tree from Carthage, in Northern Africa (whose inhabitants were called Poeni).

This is a strange, beautiful fruit containing many small seeds, each and every one surrounded by a bit of juicy red pulp. When eating pomegranates, the smashing of all of the seeds with your teeth makes a crunching noise. The tree itself is beautiful too, with gleaming green leaves and clear red flowers.

The pomegranate's juice, grenadine, is used in drinks.

A miniature pomegranate is an attractive houseplant. The fruit can be obtained during the fall and early winter. It is quite easy to get pomegranates to grow—but it is not so easy to divide the fruit in order to obtain the seeds! Usually the juice squirts out, and the stain can be hard to get out of clothing. The small seeds should be washed carefully in warm water, and allowed to dry for several days. Then put them in a glass jar, or some similar container, on a bed of moist cotton.

The seeds will grow after 4–6 weeks. As soon as a small root thread peeks out, plant the seed in a pot with regular potting soil which has been moistened. Make a small hole with a pencil about 2 cm (1 in.) deep, set the seed down into the hole with the root end down, and press the soil down. After a while, the first leaves will emerge.

Place the pot in a warm, light, and sunny spot. Do not spray the pomegranate too much, since the plant does best in relatively dry air. The conditions found in most homes should be ideal. The plant should be topped, but it is rather difficult to get it to branch, and so it is best to plant several plants together in one pot. The soil should be kept moist and it should be generously fertilized during the summer months.

After several years flowers can develop, and if you artificially pollinate them by transferring some of the pollen from one flower to another with a brush or feather, fruit can be produced.

Prickly Pear Cactus

Opuntia ficus-indica

The prickly pear cactus belongs to the *Cactaceae* family and originates, as other cacti, from America. There are over 200 different varieties of *Opuntia,* and a number of them have edible fruit. When the Spaniards came to America, they sent back to Europe many plants, among them the cactus. They did so well in the Mediterranean that they soon became wild and are now quite common there. In Australia they did even better, and the cactus turned into the worst "weed" on the continent.

The fruit varies in color from gold to red-violet, and has on its surface many glochidia, cushion-like groups of short, barbed hairs. If you skin it, which takes a certain amount of practice (be sure to handle with gloves), you will see that the pulp is pink and filled with seeds. The taste is fresh and quite good. The flowers are large, yellow, and beautiful.

The prickly pear bears several times during the year. The fruit can occasionally be bought in stores. The seeds keep their viability for quite a while, since in their natural habitats it can take a long time before the conditions are right for sprouting.

The seeds should be picked out of the fruit and rinsed carefully with warm water, after which they ought to dry. Since the small prickly pear plant is sensitive to transplanting, it is a good idea to plant the seeds from the beginning in pots about 10 cm (4 in.) in diameter. In this pot the cactus can grow for many years. You can also plant the seeds in peat pots (Jiffy pots) that swell up when watered, and the plants and pots together can later be moved into a bigger pot without requiring transplanting. This is a good method because peat is sterile, and small cacti are very sensitive to diseases of the soil.

An ideal soil mixture consists of 70 percent regular potting soil and 30 percent sand. Soil can be sterilized at home by placing it in the oven for about two hours at a temperature of 225°C. (400°F.). Then to plant the cactus, bury the seed a little way down in the soil, which should be moist.

Put the pot in a warm and light place. After about 4–6 weeks the cotyledon will come up, and this will not resemble a cactus; but later, the cactus itself will emerge from between the sections of the cotyledon. Strangely enough, two plants often arise from a single seed. One is markedly larger than the other, and it usually crowds the smaller one out until it dies. You can remove the smaller plant in the beginning so that all the energy goes to the larger plant.

The location of the little cactus should be light, sunny, and warm during the day in the summertime. Give it regular watering and some fertilizer. During the winter, place the cactus in a much cooler spot, preferably 6–8°C. (43–46°F.), if possible. During this time, watering should be lighter. After a while, flowers and fruits may develop. If you live in a warm climate, place the cactus out in a protected and sunny spot in the garden, where it can stand throughout the year.

The prickly pear plant, as a rule, has few hard thorns. Its bulbous pads, or "leaves," have clumps of barbs that easily fasten to one's clothes—and skin!

The cactus' thorns are not poisonous, but they are often dusty and so if you prick yourself, you can easily get an infection.

Quince

Cydonia oblonga

The quince belongs to the *Rosaceae* family and is a close relative of the apple and pear. It grows wild in the Near East and belongs to the group of plants that have been cultivated since ancient times in the Mediterranean. The Greeks began to

propagate the quince around 100 B.C., and it was later carried west and north to Europe. The best variety of quince comes from Cydonia on Crete, and this explains the derivation of its Latin name.

The quince used to be regarded as a symbol of luck, love, and fertility, and a bride ate the fruit to show her devotion to Aphrodite, the goddess of love.

The quince is a large bush or a small tree, and the leaves are downy on the underside. The blossoms are solitary, large, and resemble apple blossoms. The fruit is downy and resembles the apple and pear.

Quince is used in compotes, marmalade, and jam, but it is not eaten raw. In many areas the quince is quite a common ornamental bush in gardens and parks, and more people should take advantage of the opportunity to make quince marmalade, which is very good.

For the planting of quince seeds, see Apple, page 53.

Rice

Oryza sativia

Rice is another grass—the *Gramineae* family—and it is an anciently cultivated plant from southern China. From there it spread east and also west, to Europe, and rice was carried from Europe to the Americas.

Rice is the plant that feeds most people in the world. It grows up to 1 m (39 in.) in height. There is a difference between mountain rice and swamp rice. Mountain rice is grown the way other grains are grown, and it is found in most of Latin America. The cultivation of swamp rice, on the other hand, is a very complicated procedure involving the flooding of the fields. There are innumerable varieties of swamp rice.

To plant rice at home, you must use the unrefined (brown)

kind, that still retains its shell. This is available in the stores, even though most of the rice sold is refined (white).

You can either place rice in a glass jar on a wet bed of cotton and watch it grow, or plant it in regular potting soil in a pot or flat. The soil should be quite moist when planting, and a plastic bag or lid will provide high humidity. Put the pot in a warm place so that the soil is maintained at a warm temperature.

When the rice plant has grown about 10–15 cm (4–6 in.), transplant it into a pot with a tight bottom so that when you fill the pot with water, the water will remain on top of the soil. Use a clay-like soil when you transplant.

It is unlikely that the plant will be able to produce rice grains, since rice requires much sun and warmth, but it is fun to try growing it.

Rose Apple

Mespilus germanica

The rose apple belongs to the *Rosaceae* family and grows wild in Southeast Europe and the Near East. The tree can reach up to 6–8 m (20–27 ft.) high. Its leaves are perforated on the edges and somewhat downy on the underside; the blossoms are white and have short stems. The fruit is brown and open at the top, and reminds one somewhat of a rosehip. The rose apple is used for jam and jelly, but it is also edible when overripe or frostnipped.

For the planting of rose apple seeds, see Apple, page 53.

Rosehips

Rosa canina

Rosehips are the fruit of the rose bush, and the rose belongs to the *Rosaceae* family. Rosehips are an old medicinal plant and are mentioned in ancient herbal books from the fourteenth century. They are a stone fruit and contain many small seeds.

To plant rosehips, cut open the fruit and pick out the seeds. Let the seeds dry for several days. Plant them in a pot or flat with regular, moist potting soil. A plastic bag, with air holes, over a pot or a lid over a flat will increase the humidity. Water the seeds only periodically, and place the pot in a location where it gets light.

When the seedling comes up, water it regularly and put it in a very well-lit spot with protection against strong sun. Later on you can transplant it—a deep pot is recommended so that the roots can develop properly. Use an outer pot to guard against the sun drying out the soil. When the rose has grown about 10 cm (4 in.) high, give it regular nutrients or fertilizer. You should wet the leaves in the morning with a fine sprayer. Actually, the rose does not like to have water on its leaves too often because they easily become mildewed; but if they are to tolerate the dry air inside the house, they should be sprayed now and then.

Early in the summer, plant the rosebush in the garden or out in nature. After about four years, the rosehips can be harvested.

Tomato

Solanum lycopersicum

The tomato belongs to the *Solanaceae* family. The small fruit type, the cherry tomato, originally came from South America. When the tomato reached Mexico, it was seriously cultivated for the first time. The Aztecs raised tomatoes and harvested the large fruits with which we are now familiar. The name tomato comes from the Aztec.

In the middle of the sixteenth century, the tomato was brought to Europe, and there it stimulated a great deal of interest. Tomatoes were called love apples since they were thought to awaken warmer feelings in those who ate them. In the latter part of the seventeenth century, tomatoes were raised in botanical gardens in some countries but only as ornamental plants; the fruit was not eaten. Even in our century, some people have believed that tomatoes were not good for children, and it is only in the last

thirty years that the consumption of tomatoes has dramatically increased.

Some of the plants that belong to this family (deadly nightshade, for example) , are indeed poisonous, but others have the edible fruits and roots with which we are familiar, such as the tomato and potato.

Tomato seeds can be sown and will reproduce new plants. Scoop out the seeds from a tomato and rinse them well with warm water. Allow them to dry thoroughly on newspaper, and then leave them lying for several weeks in a tightly closed bag in a dry and warm place.

Plant the seeds in packaged potting soil in a pot or flat, or use peat pots (Jiffy pots) that swell up when watered. The soil should be moist when planting. Bury the seeds ½ cm (¼ in.) deep, and put a plastic bag, with air holes, or a lid over the pot or flat to increase the humidity. Place the pot in a light location, with an average temperature of 16–20°C. (60–68°F.) . Spray the top soil daily with warm water so that you do not need to water the plants as often. It is important to let air penetrate into the soil between waterings.

When the first leaves (the cotyledon) come up, move the pot to a spot that gets a lot of light, with protection from the strongest midday sun. If you have planted several seeds together in a pot or flat, the seedlings should be transplanted into their own pots when they have grown some. Subsequent transplanting should be done when the tomato plant needs more room, and you can use regular potting soil. Tomatoes do well in relatively large pots. Water regularly and generously fertilize the plant.

Many tomato plants produce side shoots and these can be cut off, so that there will be single stem growing straight up. In most cases, tomato plants must be supported with stakes or tied up.

When the danger of frost has passed, gradually accustom the plant to spending more time outside, until eventually it can stay out the entire day. The small cherry tomatoes can be grown indoors in a sunny window the entire summer, or even all year in a moderate climate. In midsummer, cut the top off the plant to prevent further vertical growth. Then all of the energy will go into the development of fruit.

Walnut

Juglans regia

Walnuts belong to the *Juglandaceae* family and grow wild across a belt of land stretching from the Balkan peninsula to the Himalayas. Walnuts are cultivated commercially in many parts of the world.

The walnut is actually a stone fruit and not a nut, but the outer meaty envelope splits and discharges the nut in its shell at maturity. The seed that lies inside the shell is mainly the embryonic cotyledon.

For the planting of walnuts, see Pecan, page 123.

Index

Bibliography

Barnens egen Trädgård, R. Pettersson, Almqvist & Wiksell, 1955
Biologi för Jordbruk, Skogsbruk och Trädgård, E. Åberg, H. Lindblom, D. Johansson.
   LT, 1972
Blomma, Frukt och Frö, B. M. Parker, A. V. Carlsson, 1962
Blommor och Träd vid Medelhavet, T. Hylander, Bonniers, 1975
Den Stora Blomboken, F. A. Novak, Folket i Bild, 1966
Den Svenska Skogen efter Istiden, L. von Post, Bonniers, 1933
Frukt och fröspridning, B. M. Parker, A. V. Carlsson, 1961
Från Kärna till Frukt, A. Mitgutsch, Bonniers, 1975
Gagnväxter, B. Jönsson, Gleerups, 1935
Kyskhetsträd och Änglatrumpeter, U. Beyron, Natur & Kultur, 1969
Myrten och Kejsarkrona, I. Lindegård, Natur & Kultur, 1971
Nyttoväxter i färg, T. Linnell, Almqvist & Wiksell, 1973
Nöjsamt, Sällsamt, Nyttosamt om Våra Växter, N. Hewe, Bergvalls, 1953
Reiseführer durch das Planzenreich der Mittelmeerländer, W. Grandjot,
   Curt Schroeders Verlag, 1962
The After-Dinner Gardening Book, R. Langer, Macmillan Publishing Co, 1969
Träd och Buskar i Europa, A. Quartier, Bonniers, 1974
Vår Gröna Jord, P. Neergaard, M. Stage, KF, 1957
Öknen Blommar, V. Täckholm, Generalstabens litografiska anstalt, 1969